make alive

—

prototypes for
responsive architectures

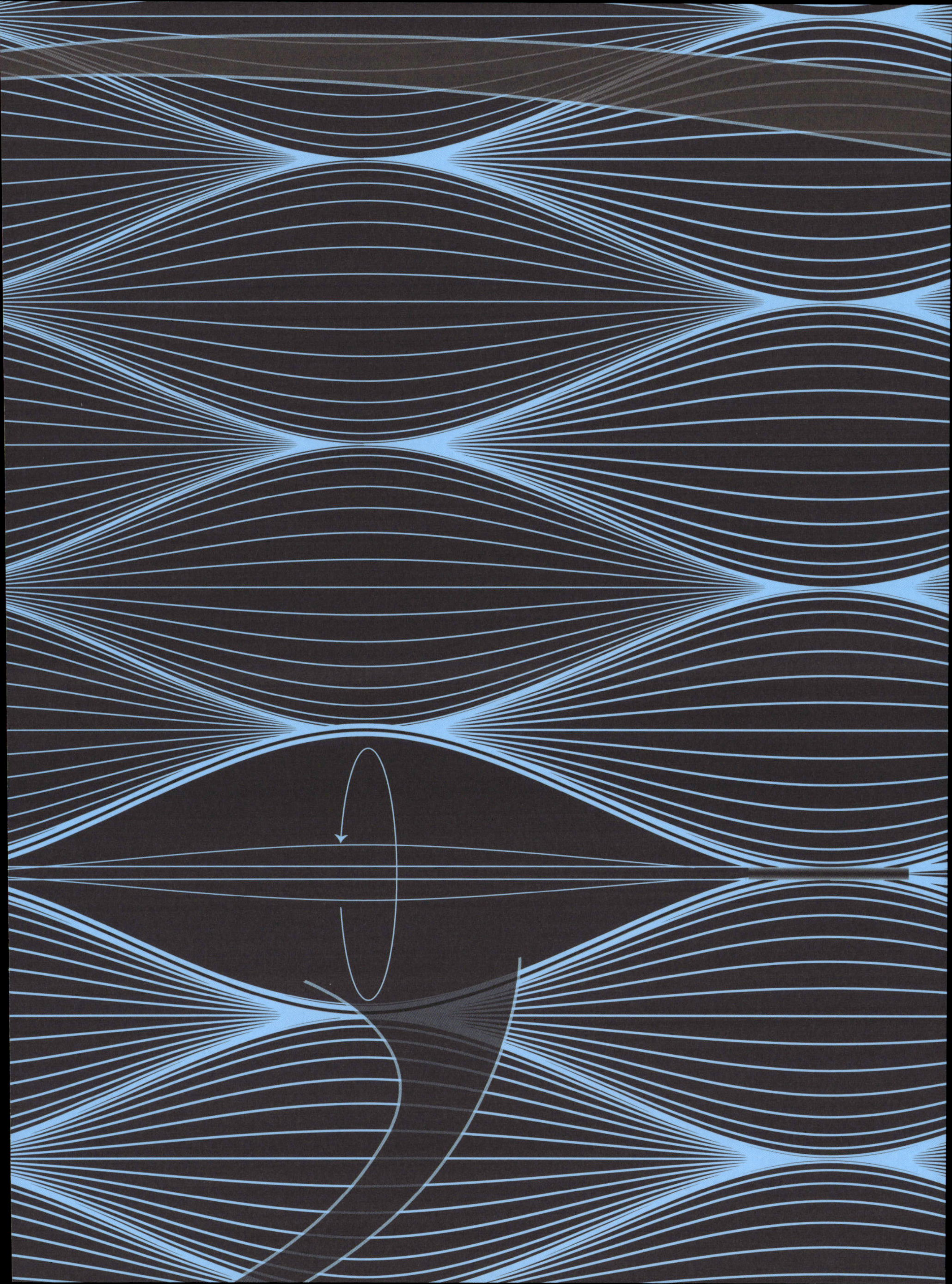

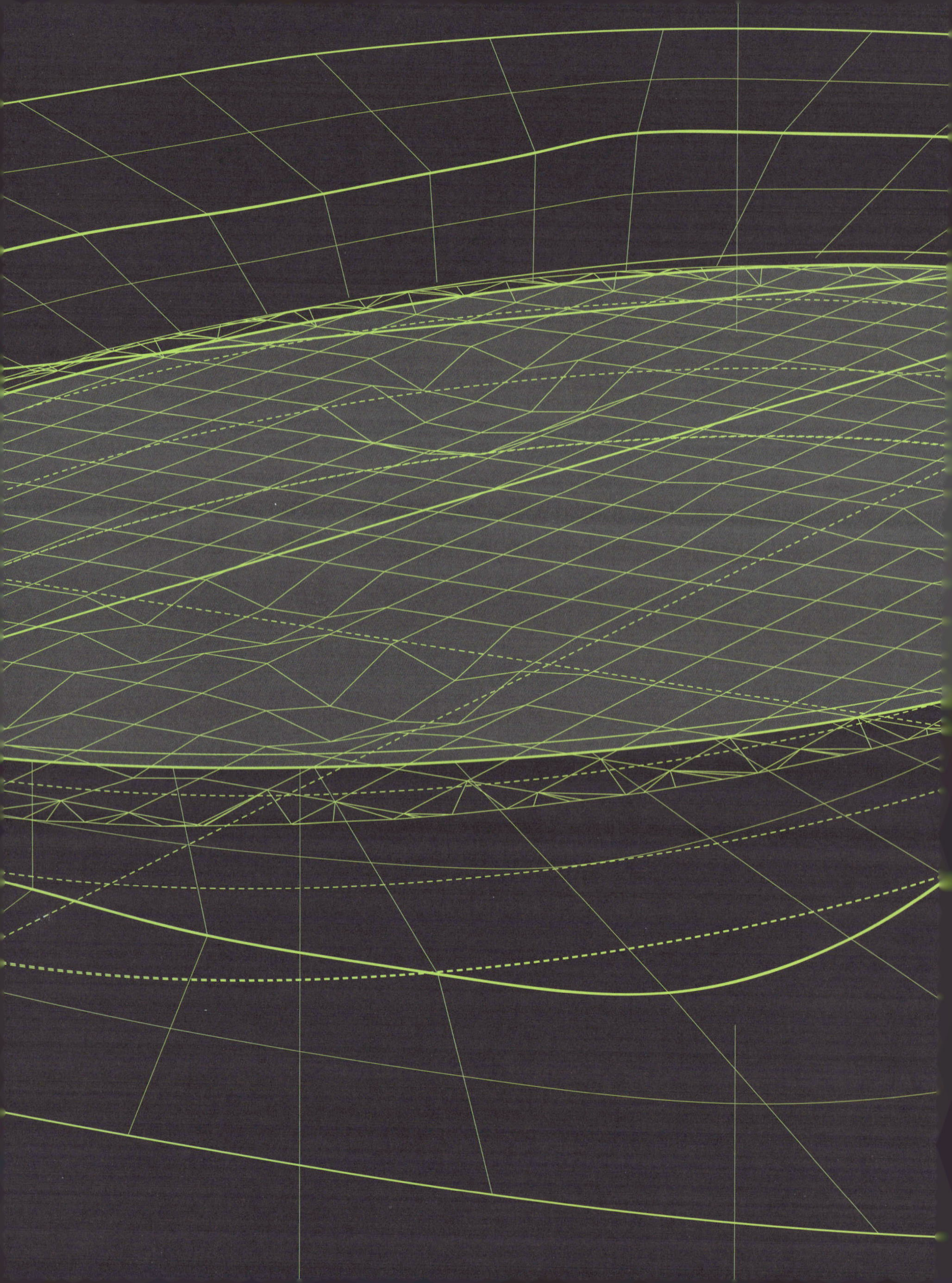

make alive

—

prototypes for
responsive architectures

Rodolphe el-Khoury
Christos Marcopoulos
Carol Moukheiber

With contributions by Drew Adams, Yuliya Bentcheva, Cleo Buster, Justin Cheng, Maya Desai, James Dixon, Holly Jordan, Khoury Levit Fong, Samuel Ko, David Long, Melissa Lui, Ricky Luk , Mani Mani, Fadi Massoud, Valentina Mele, Taraneh Meshkani, Azadeh Mohammadi, Nashid Nabian, Maggie Nelson, Otto Ng, Ada Juwah, Shadi Ramos, Julianna Sassaman, Mae Shaban, Don Shillingburg, Rick Sole, Lailee Soleimani, Studio NMinusOne, Elaine Tong, Jimmy Tran, Arthur Tseng, Nadya Volicer, and Ali Yarbakhti

OSCAR RIERA OJEDA
PUBLISHERS

contents

foreword

By Tom Verebes

"The designer is controlling the construction of control systems, and consequently design is control of control, i.e. the designer does much the same job as his system, but he operates at a higher level in the organisational hierarchy."

The Architectural Relevance of Cybernetics,
Gordon Pask, 1969

Make Alive documents an arena for design research in which distributed electronic sensors, control systems and actuators are embedded into the hard matter of architectural space, which in turn inflects to the behaviour of its interacting subjects. Through the parallel and intertwined paths of teaching, research and practice, the design work included in this book is concerned with the notion of matter as having lifelike properties, challenging the conventional assumption of material being immutable and inert. The traditional format of a book – which is limited to two dimensional and three dimensional representations and textual description, represents a challenge to capture the inherent dynamic nature of spaces and material assemblages, and their time-varying attributes, which are pursued in each project and studio included in this book.

This body of work forms part of an important arena for experimentation, one which originates with the early work in Cybernetics, and was later propelled by architects' speculations on the relationship between emerging technology and architecture since the 1960s. The cybernetician Gordon Pask wrote in 1969 that architecture need not be inflexible, and its goals have less to do with permanence and endurance, but rather, what he calls a reactive environment in which the inhabitants are in constant dialogue with the mechanisms of design or 'environmental computing machines'. He turns the role of the designer and his/her relation to the subject towards 'the interaction between the designer and the system he designs, rather than the interaction between the system and the people who inhabit it'. In this sense, the designer is no longer conceived as the authoritative controller of the final product, but rather the designer is now the designer of the apparatus which will design the product.

This book raises questions about the relationship between material and electronic technologies, and the ways in which people interact in space. The vehicles of this new paradigm rely on electronic networks and signal processing systems which enable forms of feedback between an architectural assemblage, its host environment, and its users. Through a multitude of designers, these speculative projects are less concerned with the appearance of space than with the dynamic behaviour generated through the interaction of people and the electronic compounding of physical matter. In this light, the projects and studios described in this book can be understood in their totality as experiments in spaces which perform as interfaces of multiple states. It is this heightened potential of interactivity which differentiates this body of work from the representation of conventional spatial or material projects.

Throughout the history of western architecture, the presumption that buildings should be conceived and constructed as enduring and unchanging prevails when the evidence of millennia is that this is an unattainable goal. The suppression of indeterminacy of unwarranted changes misconstrues permanence as the 'dominant cultural expectation that buildings must be built for eternity, when in fact most buildings are built to persist for only a short time.' Design is a process which aims to fix in place its configurations and mechanisms, whilst emergence 'has always been about giving up control, letting the system govern itself as much as possible, letting it learn from the footprints.' Far from making claims for a totalising ephemerality of matter, the question is why so much of contemporary architecture is seemingly so unable to adapt to the changing demands of its users and the environment? How can designers embed greater intelligence into design artefacts?

At a microscopic scale, all matter undergoes continual, albeit slow and often imperceptible, physical and chemical changes. Aside from the longer term effects of weathering on material, the ways in which architecture is instrumental to enabling specific activities also seems to lack a permanent fitness. Programmatic change is an inevitable fact which can eventually render any form dysfunctional. The projects in this book focus on the activity of users generating variable spatial configurations rather than the design of fixed, static and homeostatic architectural spaces. Far from reaching a synthesised conclusion, this book asks more questions than it answers, and thus opens a rich field of design research into intelligent and interactive environments for years to come. Given the proliferation of digital control systems into all aspects of architectural and urban space, there is the very real prospect of greater intelligence in design products with an embedded capacity to anticipate and adjust to future conditions, through memory and evaluation systems. We know the present is always, at least theoretically, in a state of flux, but Rodolphe el-Khoury, Carol Moukheiber, and Christos Marcopoulos are working on the future.

introduction

Make Alive.
Prototypes for Responsive Architectures.

The radical integration of computers into objects, buildings and cities is causing a radical transformation in the built environment. This book suggests a proactive role for architecture in shaping this new reality. The migration of computing from dedicated appliances to physical environments, thanks to increasingly proliferating microchips and ever-expanding information networks, directly empowers architecture as a transformative agent. The fact that objects can now sense, think, act and communicate with the help of embedded technology is opening up potential for an architecture that is more closely aligned with the networked dynamics of living systems – a sentient architecture. The technological enhancement of physical matter charts a movement away from a mechanical paradigm towards a biological model. This shift manifests itself on several levels: from the micro scale in the form of new composite or 'smart' materials capable of responding to external stimuli, to larger network formations between people, objects, spaces, and landscapes. Radical artifice here serves to imitate nature, enmeshing built environments in a complex web of interactions whose emergent properties approximate the resiliency of natural ecologies.

It is precisely this close synchronisation with life which has made these emerging technologies pertinent in dealing with a wide range of issues from therapeutic benefits to the body, to the mediation of global and climatic energy systems. The projects featured in this book demonstrate, through working prototypes, architectural applications of synthetic sentience in the broad research area of ambient intelligence, responsive and interactive systems, augmented reality, embedded/situated technology, mobile computing and locative media.

The research presented in this book is premised on the notion that every building or landscape component can be equipped with computational power. It documents multiple research trajectories launched and guided by the authors at The University of Toronto, MIT and the University of Hong Kong. The projects have established an interdisciplinary platform involving artists, designers, scientists and engineers and an approach to spatial problems that is attuned to the dynamics of living systems. The collective aim is to develop a digitally enhanced architecture from the collaborative experiments that is well equipped to handle persistent and emerging challenges.

Embedded technologies and sentience

"Ubiquitous computing is roughly the opposite of virtual reality. Where virtual reality puts people inside a computer-generated world, ubiquitous computing forces the computer to live out here in the world with people." Mark Weiser

The story is by now well known: 'embedded,' 'responsive,' 'adaptive,' 'situated,' 'ambient,' and 'interactive' have become more or less synonymous terms to describe a condition of 'ubiquitous computing'. This term was coined by Mark Weiser at Xerox PARC in the 1980s to describe the migration of computing power from personal computers into the physical environment – a development made possible by the evolution and proliferation of small, cheap sensors and microcontrollers. Weiser described this new emerging reality of pervasive technology as the third wave in computing, where the first wave brought forth the development of the mainframe computer, and the second the personal computer.[1] The process of continuous miniaturisation and dispersion of ever-growing computing power is becoming increasingly evident in the objects that surround us. From the autonomous robot vacuum that cleans the house when no one is home, to the coffee shop that can track a customer's location in the city through a phone's embedded GPS sensors, new affordable technologies endow previously inert objects with the capacity to sense, process and act upon incoming information. Life is infused.

From an organicist's point of view the vibrancy of life does not manifest itself simply in organic matter; it extends throughout the entire universe dissolving distinctions between organic and inorganic matter. All form, natural or artificial, can be said to emerge from the processing, or the computation of multiple form-shaping pressures. The suburban detached house typology, for example, is the result of a particular entanglement of dynamic socio-economic and technological processes that have a specific morphological manifestation. Inert as it may seem, the single family house is alive. The desire to harness and maximise the affective qualities of an object – to animate the seemingly static – has a long and diverse lineage in the history of architecture: from the immersive qualities of Islamic patterning to the plasticity of matter in Baroque architecture, to more contemporary experiments in digital fabrication and machining in extending the expression of material properties. Parallel to this understanding is the development of synthetic sentience.

Synthetic sentience

The physical world's emerging synthetic sentience mirrors the historical trajectory of increasingly decentralised digital information processing and communication systems. The 20th century – driven in part by the evolutionary forces of the US military-industrial complex, and its need to gather and manage information streaming to and from its hot and cold war battlefields – witnessed a shift from centralised and sequential or linear information processing systems to systems based on parallel computer architectures. This enabled the processing and filtration of multiple, disparate and simultaneous sources of information in recognition of patterns, or coherence, resembling more closely the complex and intuitive processes of human cognition and decision making. The notion of a 'distributed rationality', one that relies on information gathered and relayed from the environment and from the immediacy of a situation, could be integrated and processed into a 'decision' – an emergent phenomenon – in a much more fluid way, bypassing the lag in response time and detachment of a top-down command structure. This continuous drive to approximate biological intelligence has had a transformative effect on the development of Artificial Intelligence systems. The ultimate model of such a distributed system is of course the network of brain neurons.[2]

Composed of billions of neurons, the brain serves as a primary model in the evolution of artificial intelligence. The brain has no command centre, yet is able – through its low capacity but highly entangled neurons – to perform qualitative actions more effectively and faster than any digital computer to date. This ability has been attributed to the speculation that neurons function as quasi-autonomous agents with independent goals, seeking attention in a cacophonous and competitive environment. Intelligence, let's say as manifested in the act of speech, is seen then as emergent cooperation born out of this intense competition between various agencies, not unlike the objectives of parallel based architectures.[3]

Architecture as a collective of sentient agents

One of the first to explore and contribute to what is commonly referred to as the new AI, or a bottom up approach to AI was Rodney Brooks. Recalling Valentino Braitenberg's *Vehicles: Explorations in Synthetic Psychology*, Brooks' autonomous robots were conceived as agents, literally as modules, each with its own set of basic rules for perceptual and sensory-motor actions, relieving the robots from having to carry 'a world map' in the form of an overly complex centralised communication and memory system. Each robot, equipped with a closely coupled system of sensors and actuators, functioned much as an insect, sensing and acting upon its immediate environment only. This intimate interaction with the world recognises the world itself as the bearer of complex distributed and streaming information. Cognition here emerges through this situated and embodied interaction.

In contradistinction to the predetermined programmed smart house, the projects sampled in this book demonstrate patterns of synthetic sentience that emerge from a network of situated interaction: a meshwork of sensing, communicating and actuating agents embedded in all elements of the built environment. Yielding new and unanticipated behaviours and forms, these prototypes provide a glimpse into architectures yet to come.

Notes:

1 Weiser, Mark, "The Computer for the 21st Century", *Scientific American Special Issue on Communications, Computers and Networks*, September, 1991

2 Hookway, Branden, *Pandemonium: The Rise of Predatory Locales in the Postwar World*. New York: Princeton Architectural Press, 2000

3 Kelly, Kevin, *Out of Control: The New Biology of Machines, Social Systems and the Economic World*. Boston: Addison Wesley, 1994

artificial nature

The highly evolved bio-chemical processes of the vegetal species are enlisted into the mechanism of environmental control. The incorporation of organic matter into mechanical systems and its integration with the modular logic of construction produces a hybrid artefact. A new species emerges: a plant cyborg – part living, part technological – working symbiotically on a chemical level with outputs and inputs from the environment and the user. By putting nature to work, these projects produce energy, purify air, sequester carbon, and remediate landscapes. As in biological systems, resilience here is defined by a heterogeneity of species working in a productive ecosystem.

filtration block

A modular structural unit that performs as an indoor air filtration system / **Elaine Tong***

Filtration Block is a modular structural unit that performs as an indoor air filtration system to create healthier air by absorbing common indoor air toxins. The Filtration Block module utilises common indoor plants as filters.

The plants' roots are highly efficient at absorbing an array of indoor chemicals, such as Formaldehyde and Benzene. The module structure is based on the Weaire-Phelan geometry for maximum spatial flexibility. The modules lock together to form a structural wall or ceiling system. Each module is glazed for light exposure, and acts as a micro-greenhouse. The plants are sustained by a water misting infrastructure that calibrates the precise delivery of water through the use of humidity sensors and atomisers.

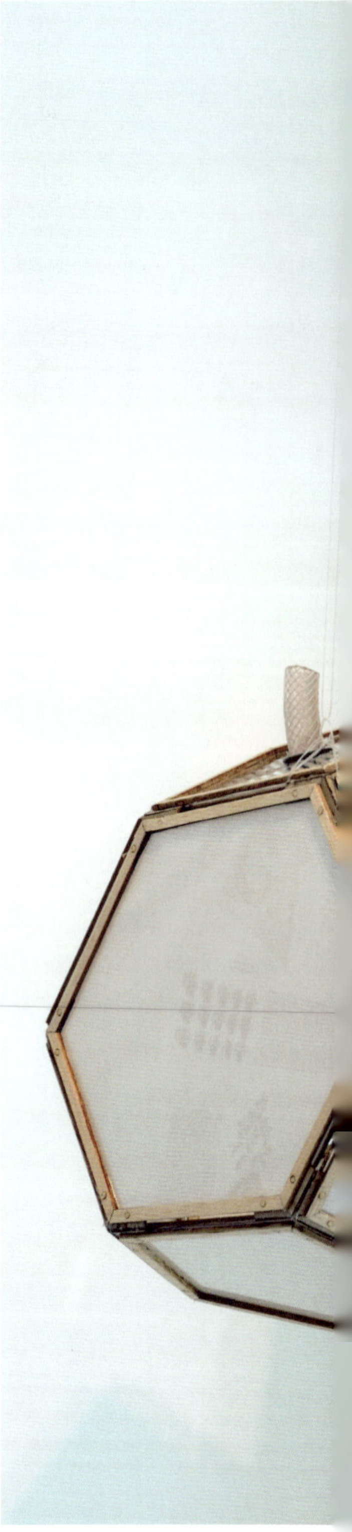

/01
Prototype showing an aggregation of modules. The opaque modules encapsulate the roots, carrying air intake and misting infrastructure. The transparent modules enclose the ornamental aspect of the plant.

*Guided by Christos Marcopoulos and Carol Moukheiber

/01

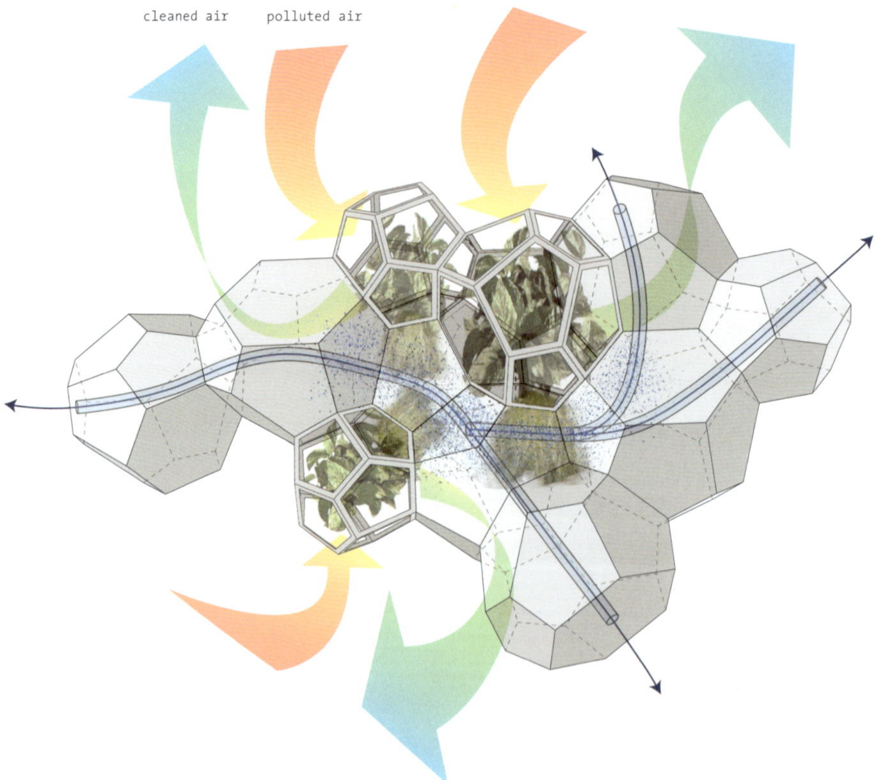

/02

/03

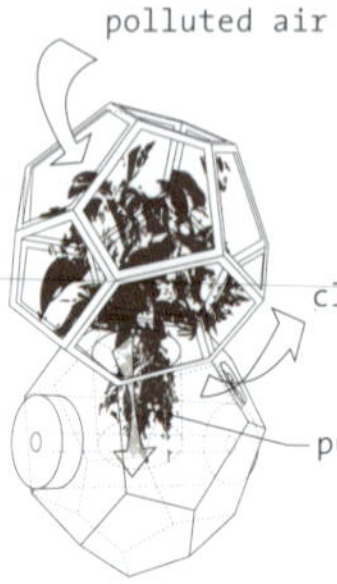

/02
Polluted indoor air is pulled in through the root-carrying modules. The air is filtered by the roots, and then released back into the atmosphere.

/03
Leaves vs. roots.

/04
Various plants are used for their ability to break down specific toxins. Some plants are optimal for the filtration of one chemical, while others filter several common indoor toxins.

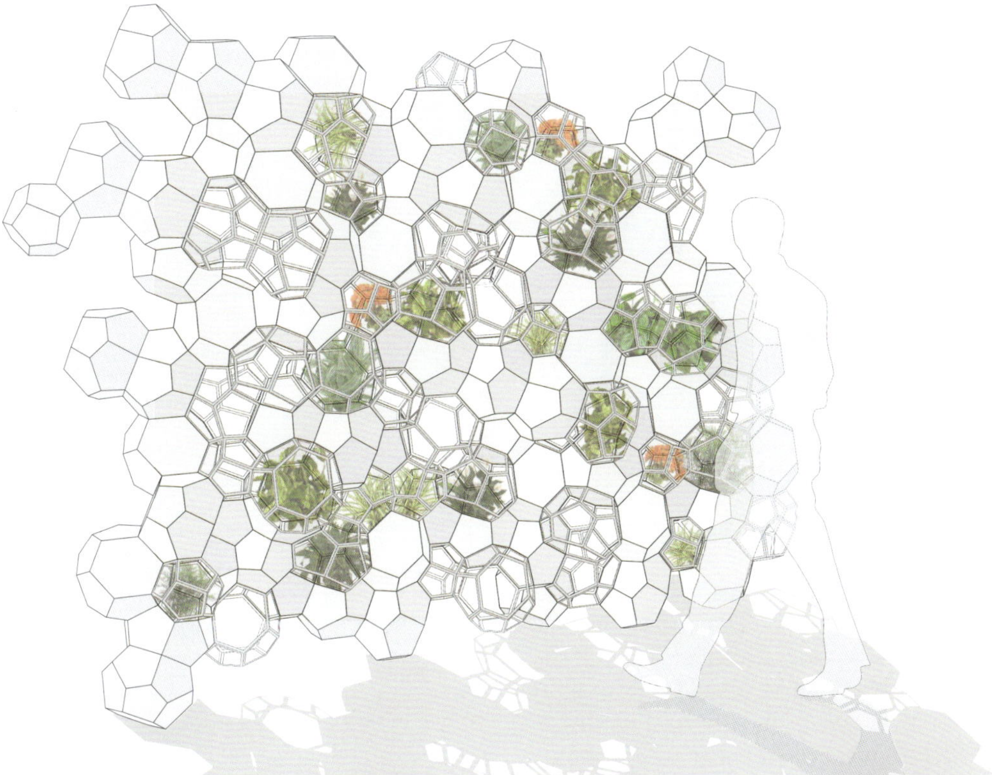

Selloum Philodendron
(Philodendron
Bipinnatifidum)

Warnek Dracaena
(Dracaena Deremensis
Warneckii)

Red Edged Dracaena
(Dracaena Marginata)

Pot Mums or
Florists Chrysan
(Chrysantheium
Morifolium)

Elephant Ear
Philodendron
(Philodendron
Domesticum)

Weeping Fig
(Ficus Benjamina)

Cornstalk Dracaena
(Dracaena Fragrans)

English Ivy
(Hedera Helix)

trichloroethylene
formaldehyde
benzene

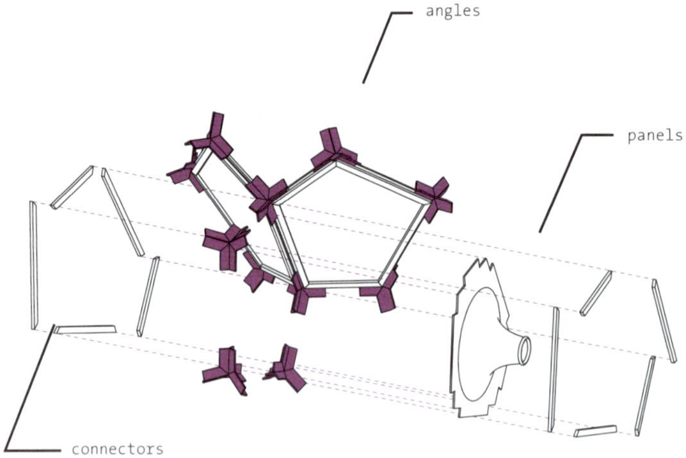

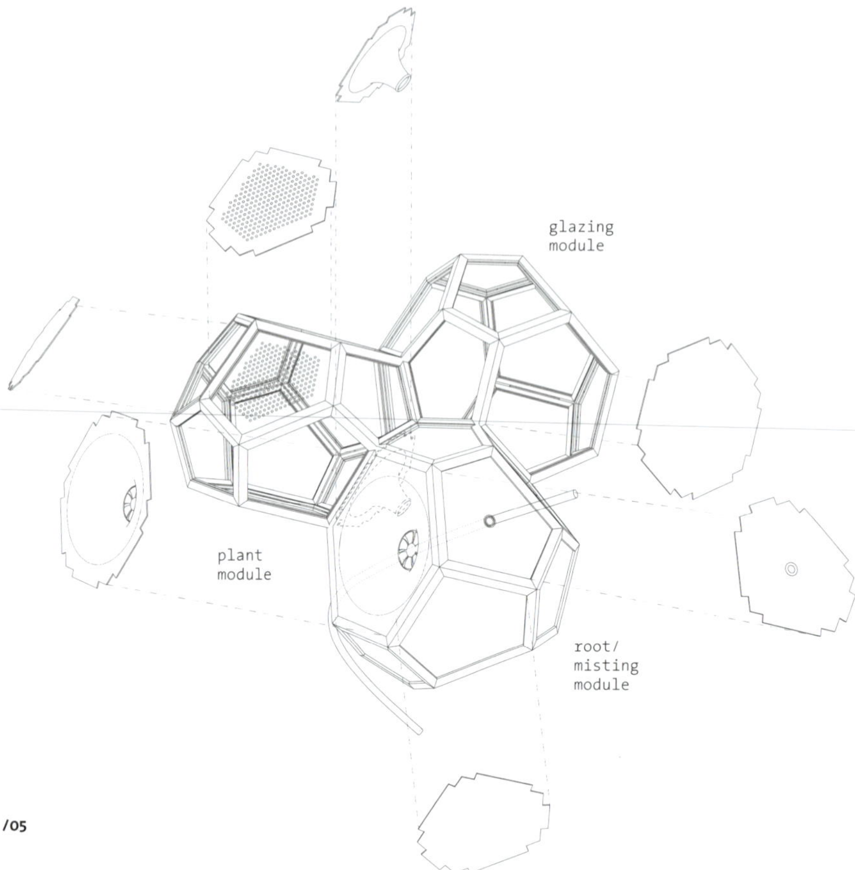

/05

/05
Anatomy of a module. The module comes as a kit of parts, and can be assembled through a system of connectors. The modules can lock together into different configurations, ensuring structural integrity.

/06
The ridged edges of the frames snap onto one another and hold the opaque and translucent panel system in place.

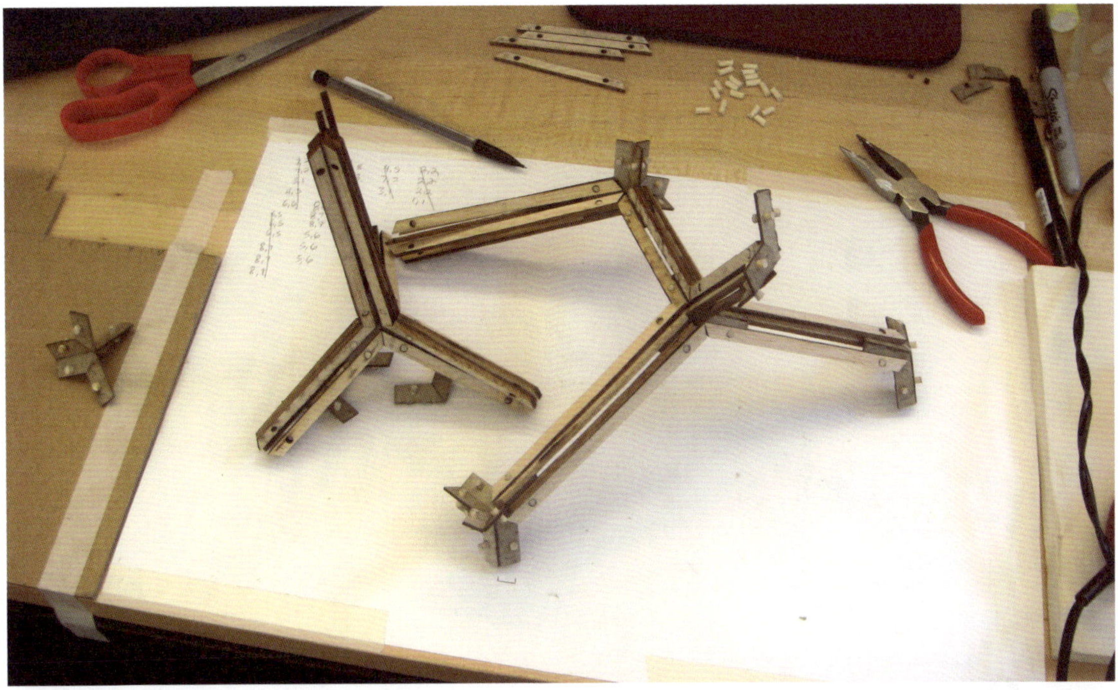

/06

MOCAPE

A bio-reactor integrated into the building envelope gives a striking leafy character to the façade while greatly enhancing the building's environmental performance / **Khoury Levit Fong**

This project for a 75,000 square metre museum of contemporary art and exhibition hall for the city of Shenzhen was one of four finalists in a combined open and invited international competition.

The project is simultaneously monumental and small in scale. It establishes a clear and elemental figure at the scale of the civic mall that it faces, while the interior breaks down into numerous complementary museum volumes and outdoor courtyards. These courtyards are open and accessible to the surrounding city and shelter a finer tuned public scale that is an alternative to the grandeur of the mall.

The exterior of the building combines two languages: mineral and vegetal. The crystals are enclosed in dark glass of varying transparency and set within a structural frame of angled members. Materially and geometrically the surface and structure of the crystals reinforce its cool mineral identity. The 'bridges' which join the crystals and close the perimeter figure of the building are covered in curved algae-filled tubes and supported by a branching pattern of columns. The bridges are arboreal in nature and create thresholds onto the landscaped courtyards within the building. The algae tubes are an innovative ecological system that conditions the building's air and produce biomass used as an alternative energy source. The patterning of the algae tubes on the façades creates a vegetal iconography from biological machinery. These tubes, seen through the leafy bamboo that grows from the courtyards and sidewalks, produce an alluring affective garden in the city.

/01

/01
Bird's eye view of MOCAPE.

/02
Detail of bio-reactor and elevation.

/02

/03
Algae culture in bio-reactors.

/04
Diagram and detail of bio-reactor.

/05
Tree-like bridging block with leaf-like bio-reactors.

/06
Bird's eye view of elevation along the civic mall.

/07
Study of light as filtered through the bio-reactors.

/03

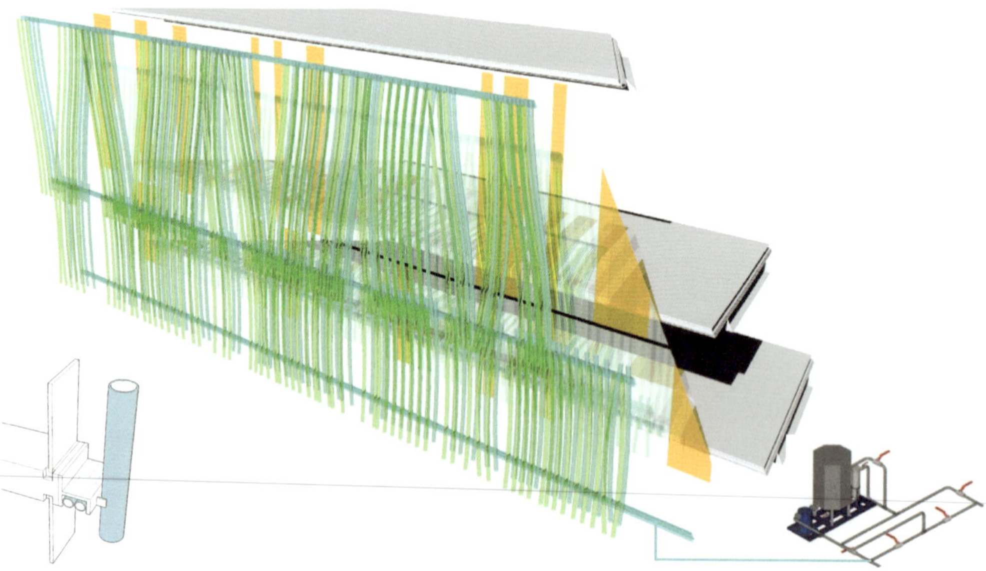

/04

/05

/06

/07

photosynthetic façade

Injecting the building's skin with an algae production farm, a building envelope aims to become a net producer of energy / **Holly Jordan***

Photosynthetic Façade is an envelope system suited to the high demands of fluctuating climatic regions. It uses embedded algae production to create a high performance envelope system with minimal net heat loss and gain.

This proposal focuses on the potential found within building skins and their unique set of functional criteria. The envelope system modulates an occupant's experience and exposure to exterior elements. Sustainability objectives focus on reducing energy use (and energy waste), thus placing even greater demand on high performance building envelopes. In short, we want maximum views to the outside while we sit in the modulated comfort of an interior space, paying as little as possible to achieve this.

For as long as our internal environmental temperatures differ from external temperatures, net heat loss or gain is inevitable. We therefore need to develop alternate means to offset that loss or gain. By thickening the building skin and injecting it with an algae production farm, this building envelope aims to become a net producer of energy as opposed to the 'standard' mediated heat loss/gain system. A network of flexible, ETFE tubing houses the algae, suspended in water and placed inside a double skin façade system. The internal air naturally stratifies and exhausts at the ceiling of each floor into the double skin cavity. The CO_2 laden return air feeds a series of bubblers which aerate the algae tube system. In conjunction with its exposure to the sunlight, the algae feeds on the CO_2 and therefore grows. The water solution within the tubes also acts as a heat sink to absorb heat from sunlight and exhaust air. The heat energy can be redistributed to other zones of the building by passing air through the double skin cavity. Over time, the algae solution intensifies in colour as it develops, until

it is harvested through a main drainage system. The cycle is completed once the tubes are refilled with fresh, young algae solution. In response to occupant use, activity, and outward expression, the tubular system can be manipulated and controlled by a system of muscle wire. Muscle wire responds to charges sent through it by elongating or shrinking. The resulting network is a curtain of tubes surrounding the building.

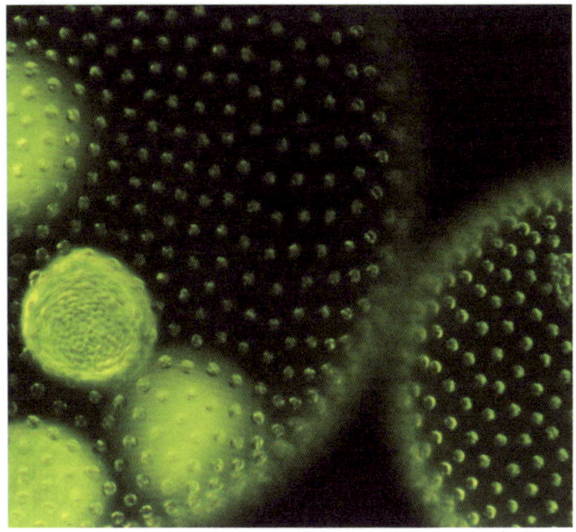

*Supervised by Rodolphe el-Khoury

/01
Elevation and section of façade integrated bioreactors.

/02
Plastic tubes deform by means of muscle wires in response to climate and the presence of viewers. The field thickens to shield from direct sunlight or loosens up to yield the view.

/03
The plastic tubes of algae culture extend along the roof to provide shade and most efficiently capture the energy of the sun.

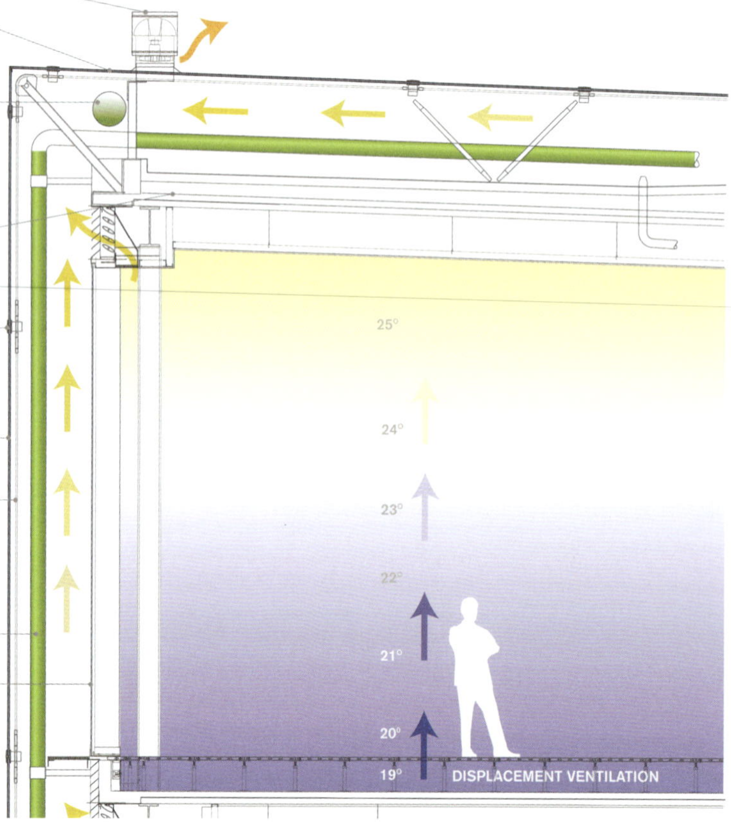

/01

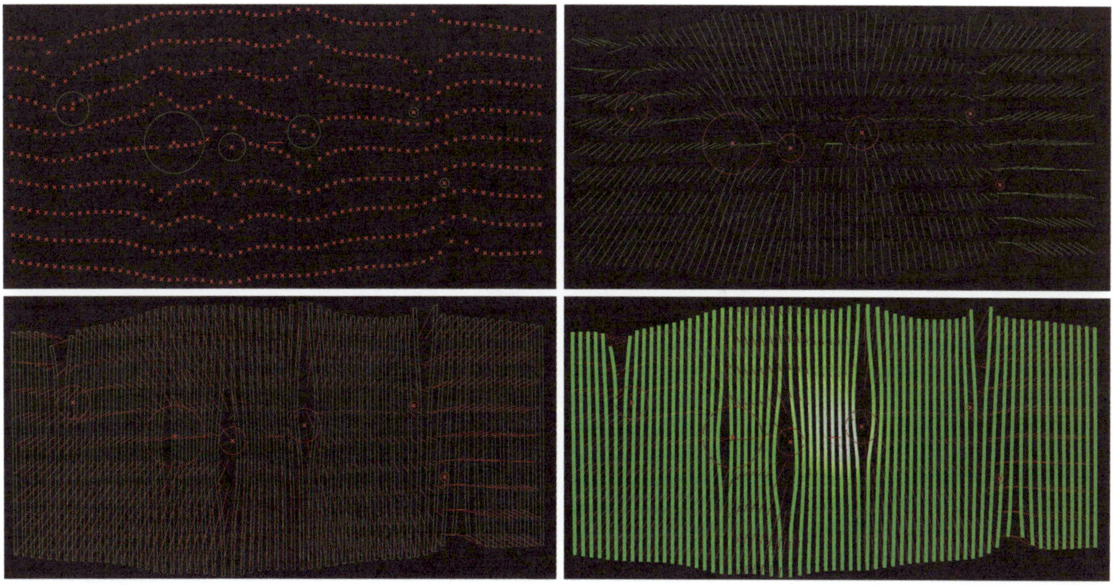

/02

/03

oxygen curtain

An algae photosynthetic gas exchanger takes the form of a curtain, oxygenating indoor air while modulating light / **Mae Shaban***

The Oxygen Curtain integrates an efficient organic living carbon sink into an interior space, responding automatically to fluctuating CO2 levels.

The curtain integrates an efficient organic living carbon sink into an interior space. The curtain produces an amount of oxygen equivalent to a mature broad-leaved tree – it is a dramatically enhanced house plant. The curtain is composed of an array of algae bioreactors. A network of indoor air, power and nutrient supply lines weave the bioreactors into a single membrane. The nutrients are supplied by the building's waste water. The curtain is nourished by the CO2 exhaled by the inhabitants. It is directly responsive to the users and the environment; each module operates autonomously and sensors activate select modules as appropriate to the changing levels of CO2 within a space. The modules then expand and contract with circulating air revealing a mechanic-organic organism that is continuously refreshing the air.

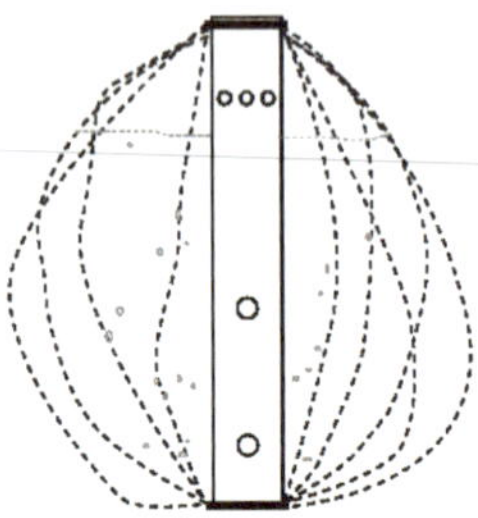

Passive Growth Module

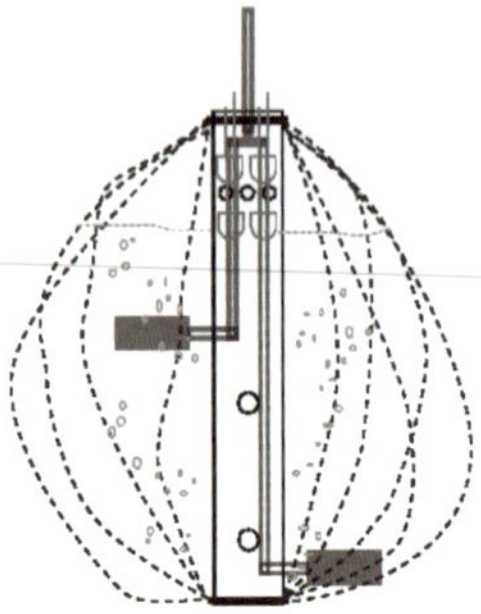

Active Growth Module

/01

/01
The curtain contains two different types of bioreactors: the active growth modules are equipped with aerators and LED lights that speed up the growth cycle of the algae. The passive growth modules rely solely on ambient light within a space. The curtain exists in mixtures of active and passive modules.

/02
Electrical and aeration infrastructure supply tubes form the structural matrix for the algae bags.

*Guided by Christos Marcopoulos and Carol Moukheiber

/02

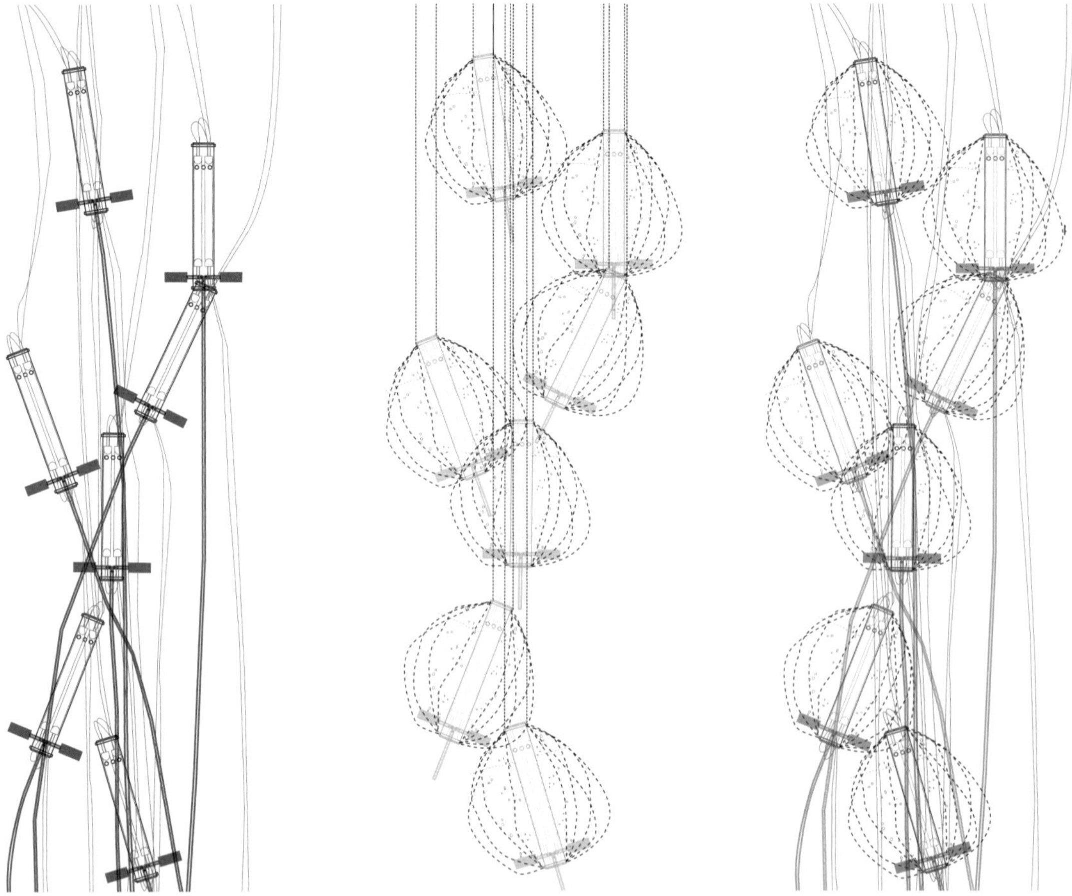

Electrical and Ventilation

Structural Wiring

Curtain

/03

/03
Each algae bag grows at a slightly different rate, creating different algae densities within the bags. The modulation of natural light becomes a function of changing algae density.

synthetic landscapes

A case for the implementation of a Beta landscape at the Gaspé Mines, Québec / **Maya Desai***

Challenging the notion of 'naturalness' and presenting a new means of remediating irreversibly degraded land areas, Synthetic Landscapes deploys a range of cybernetic mechanisms to create an alternative ecosystem that mimics the performative and experiential qualities of its natural counterparts.

Situated at the Gaspé Mines near Murdochville, the proposed project is a synthetic provincial ecotourism park. Drawing on research from across various fields of study, from synthetic biology to robotics, this project brings together seemingly opposite notions of nature and machinery. Based on automated systems, the implementation of synthetic and interdependent hydrological and energy cycles creates the rudimentary ecosystem for the Synthetic Landscape, while designed elements of performance, interactivity, and programmed environmental cues create the experience of nature. Tourists and researchers alike are invited to camp in the park, experience their surroundings, and conduct their own experiments at various scales, thus contributing to a constantly evolving, synthetic nature.

Four classes of technological interventions constitute the backbone of the proposed synthetic landscape: automated pollutant collectors with embedded swarming logic constantly navigate the site, creating a dynamic and intelligent maintenance system. Cyber bugs – flying robots that monitor the landscape's concentration of toxic gases – provide visitors with visual clues about environmentally safe zones for camping, simultaneously communicating this information wirelessly to the swarming collectors. The photokinetic trees mimic the experiential aspect of natural trees while functioning as collectors of natural energy and producing hydrogen from the water harvested onsite. Rainwater catchment ponds retain and remediate the rainwater used onsite both for camping and hydrogen production.

/01

/01-02
Artificial earth-scrubbing insects.

*Supervised by Rodolphe el-Khoury

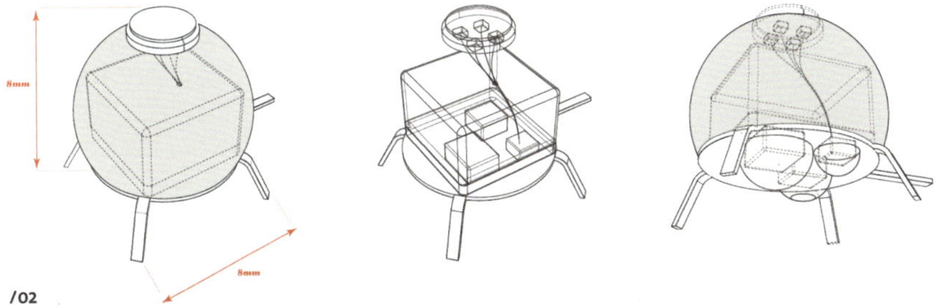

/02

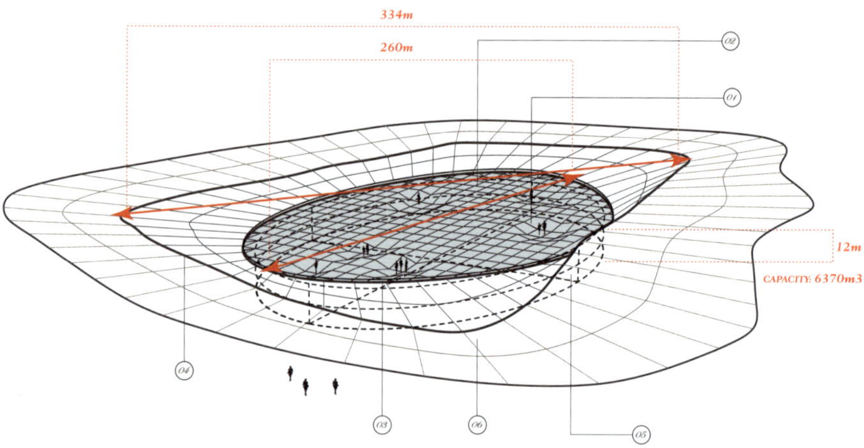

01 PVC-COATED POLYESTER FABRIC WITH THIN FILM PHOTOVOLTAIC COATING:
Serves as protective, flame-retardent, high tenacity impermeable top layer w. solar potential for cistern pump system.

02 FLEXIBLE WOVEN DRAWN CARBON STEEL ANTICLASTIC MEMBRANE (20MM CABLES):
Preserves equilibrium & keeps structure stable, while allowing flexibility & deformation.

03 PERIMETER STEEL SUPPORT STRUCTURE:
Stabilizes steel membrane.

04 INFILTRATION DUCTS:
Allows for perimeter drainage.

05 CIRCULATION RESERVOIR:
Serves as a water reservoir for supply tanks & campsite utilities.

06 CEMENT MODIFIED SOIL:
Serves as a structural base while retaining soil properties.

/03

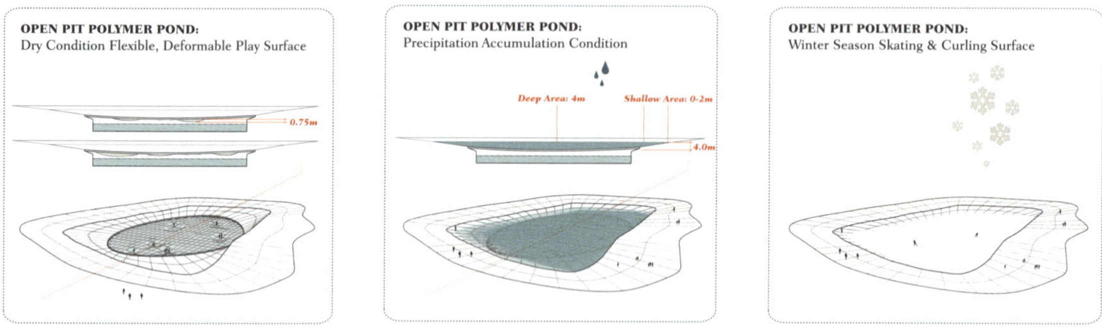

/04

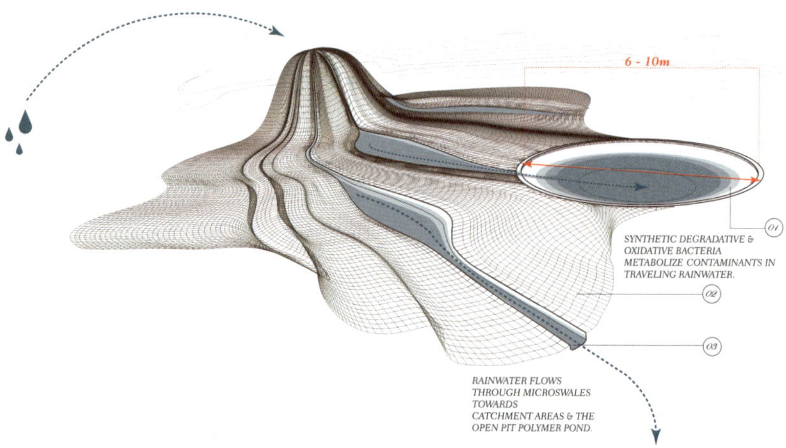

01 RAINWATER CATCHMENT PONDS:
1. Temporarily retain & remediate rainwater surface runoff.
2. Supply water to rest areas & to photokinetic trees for hydrogen production.

02 BIOPOLYETHYLENE SOIL SCAFFOLD:
Prevents topsoil erosion while allowing water penetration & provides stability on sloped areas.

03 BIOLUMINESCENT BACTERIAL REMEDIATION MICROSWALES:
Synthetic degradative & oxidative bacteria metabolize contaminants, emitting light when active.

/05

/06

/03
Sealed water treatment reservoir.

/04
Polymer seal of reservoir in different states.

/05
Synthetic degrading and oxidating bacteria metabolise contaminants in travelling rainwater; rainwater flows through micro-swales towards catchment areas and the open pit polymer pond.

/06-07
Artificial chemical-sniffing insects.

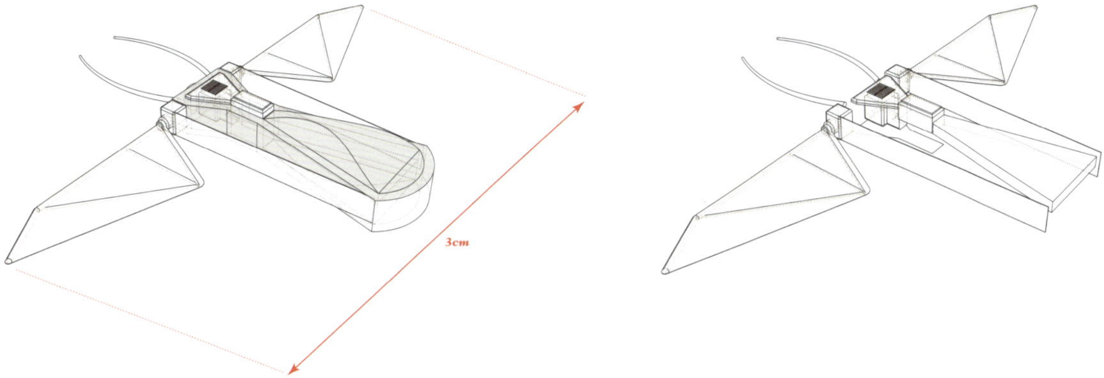

/07

/08

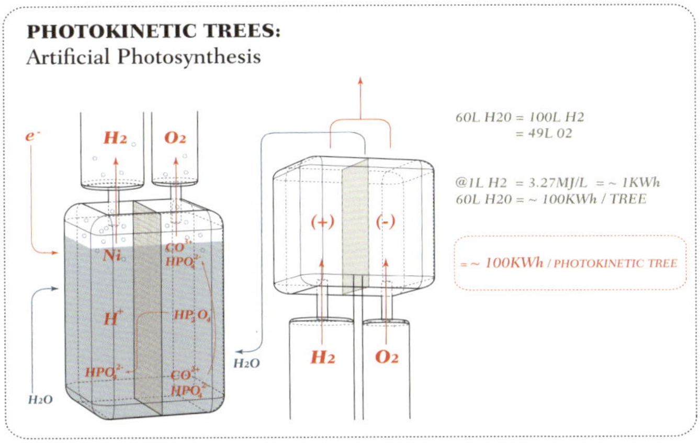

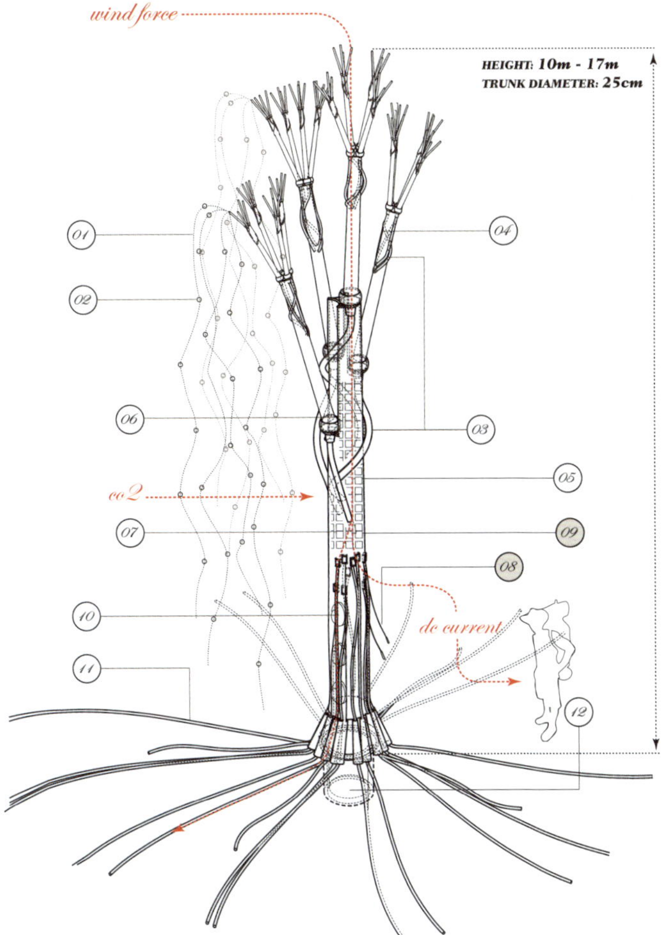

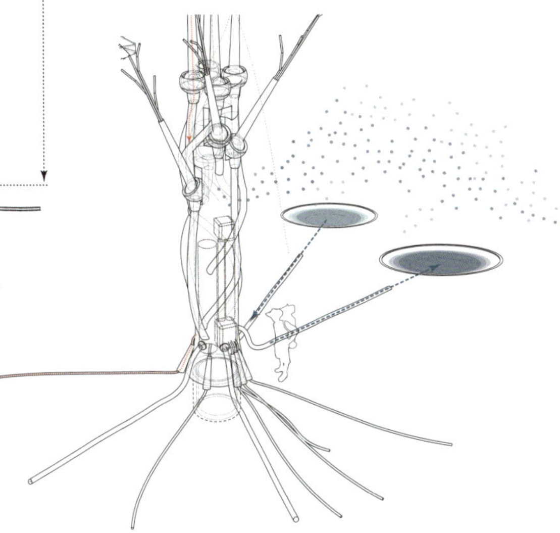

/08
Views of the artificial landscape.

/09
Diagram of a synthetic tree showing energy-generating leaf-like piezoelectric filaments.

01 **PIEZOELECTRIC POLYVINYLIDENE FLUORIDE (PVDF) BRANCHES.**

02 **POLYMER FILAMENTS**

03 **AMBERSTRAND WIRES ENCASED IN ZYLON POLYMER.**

04 **BIODEGRADABLE CARBON POLYMER BRANCHES**

05 **CARBON POLYMER TRUNKS.**

06 **LIQUID HYDROXIDE CO2 FILTERS**

07 **CARBON COMPRESSION STORAGE TANK.** (650L)

08 **SOLAR RE-CHARGEABLE LIGHT FILM.** (8hr Duration)

09 **OUTLET** (Low Volt Applications)

10 **CARBON RELEASE VALVES**

11 **NETWORKED ROOT SYSTEM** (Amberstrand Wires Encased In Zylon Polymer).

12 **CASON.**

/09

the out-house

The yard becomes an organic machine for the harvesting and filtration of a single body of water that is perpetually recycled in a semi-closed system / **Studio NMinusOne with Don Shillingburg**

The project proposes an inverted McMansion in an arid climate which, in its final act of over-consumption, ingests its own rear yard. However this act leads to its own survival. The ingested yard now becomes an activated organic machine for the harvesting and filtration of a single body of water that is perpetually recycled in a semi-closed system.

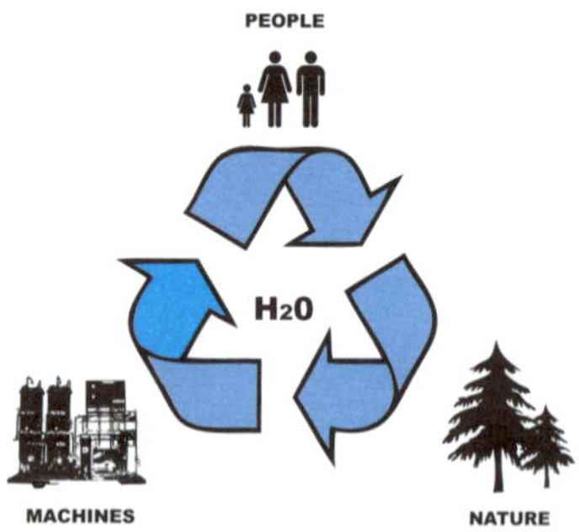

The original pitched roofs made for shedding water are now opportunistically reconfigured to collect water, and harvest energy. The newly ingested yard/lawn becomes a lush living filtration system cleaning grey and black water for reuse, while offering an Eden-like garden for its inhabitants. Eden is sustained by the toilet; a site for human waste disposal becomes a site for production, providing the necessary nutrients for the maintenance of plant and animal life, which in turn clean the water.

What was once the exterior of the house is now the most private area. The most private moments are experienced in what used to be outside.

In 2004, after five years of severe drought, Las Vegas banned all lawns in any new housing development, allowing 'desert landscaping' only. Residents are paid $1 for every square foot of lawn they dig up.

"Long term, we have to keep growing, but it is going to be a different growth. To put it bluntly, in this town we are going to drink what we flush."
Oscar Goodman, Mayor of Las Vegas

/01
The McMansion is quartered and inverted on its site. The yard now occupies the centre.

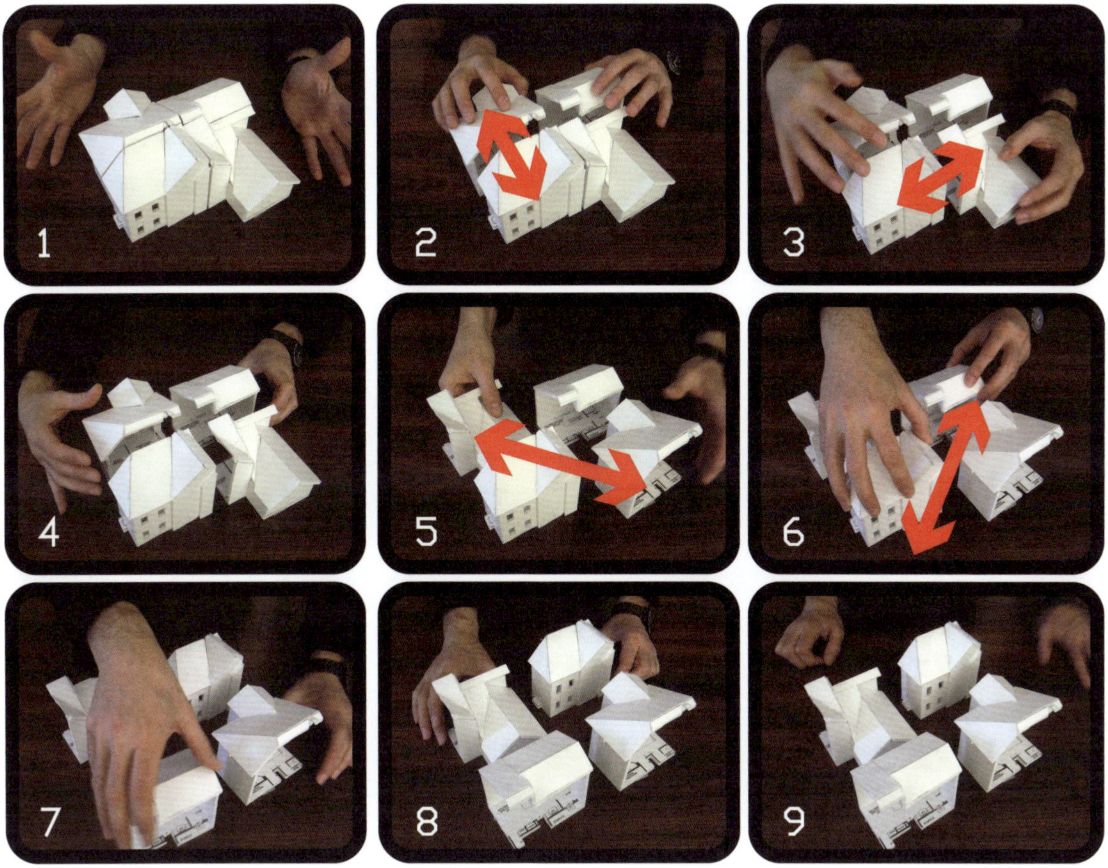

/01

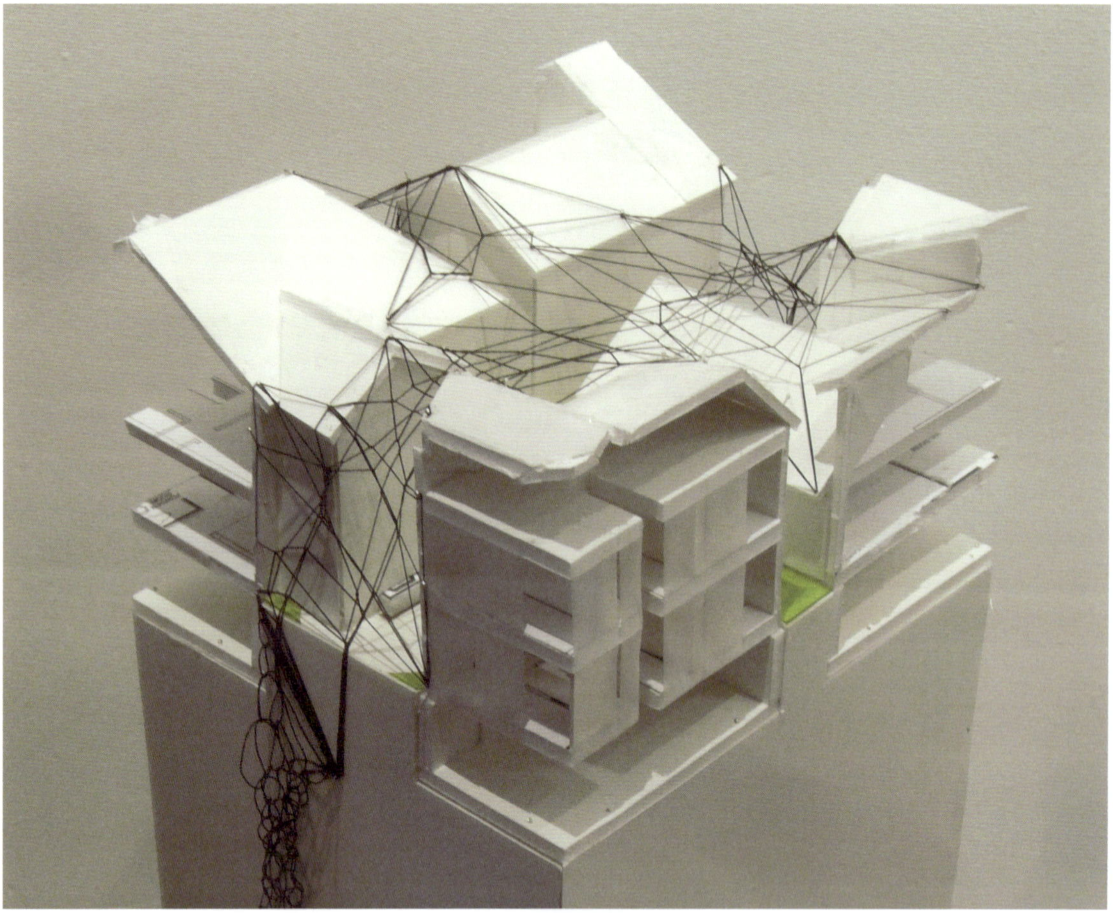

/02

/03

/04

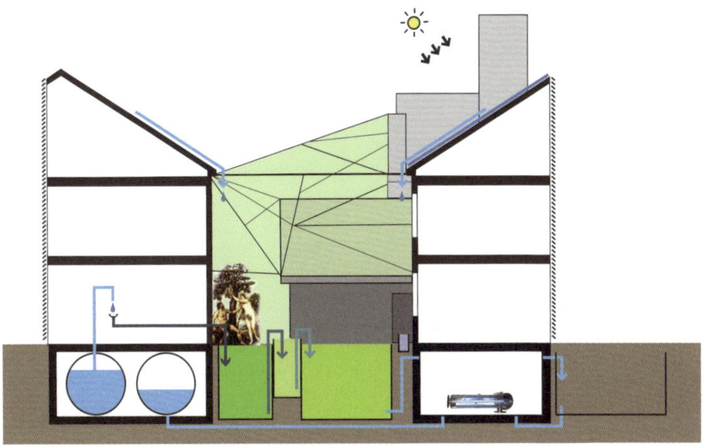

/05

/02
The interior garden is covered by the net, an impermeable membrane that prevents water evaporation. Water condenses on the interior surface, dripping back into the garden.

/03
Through its reconfiguration/ inversion, the McMansion's form evolves towards a more contemporary state.

/04
The yard is composed of a series of tanks that make up the living system machine.

/05
The section shows the cycle of water usage: from the basement storage tank to the wet zones (kitchen and bathrooms), into the living filtration system, through a UV filter and back into the tank.

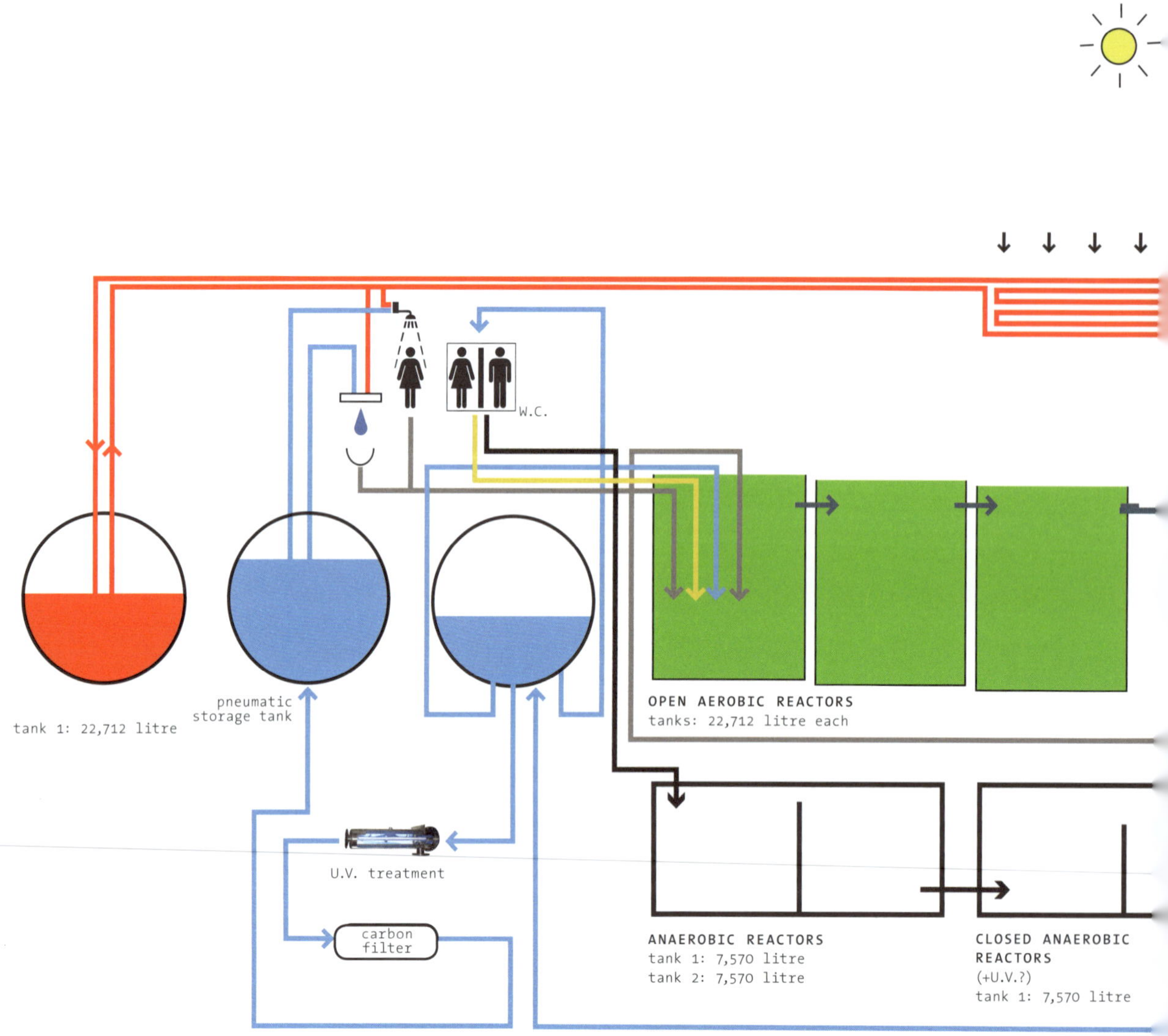
W.C.
tank 1: 22,712 litre
pneumatic
storage tank
U.V. treatment
carbon
filter
OPEN AEROBIC REACTORS
tanks: 22,712 litre each
ANAEROBIC REACTORS
tank 1: 7,570 litre
tank 2: 7,570 litre
CLOSED ANAEROBIC
REACTORS
(+U.V.?)
tank 1: 7,570 litre

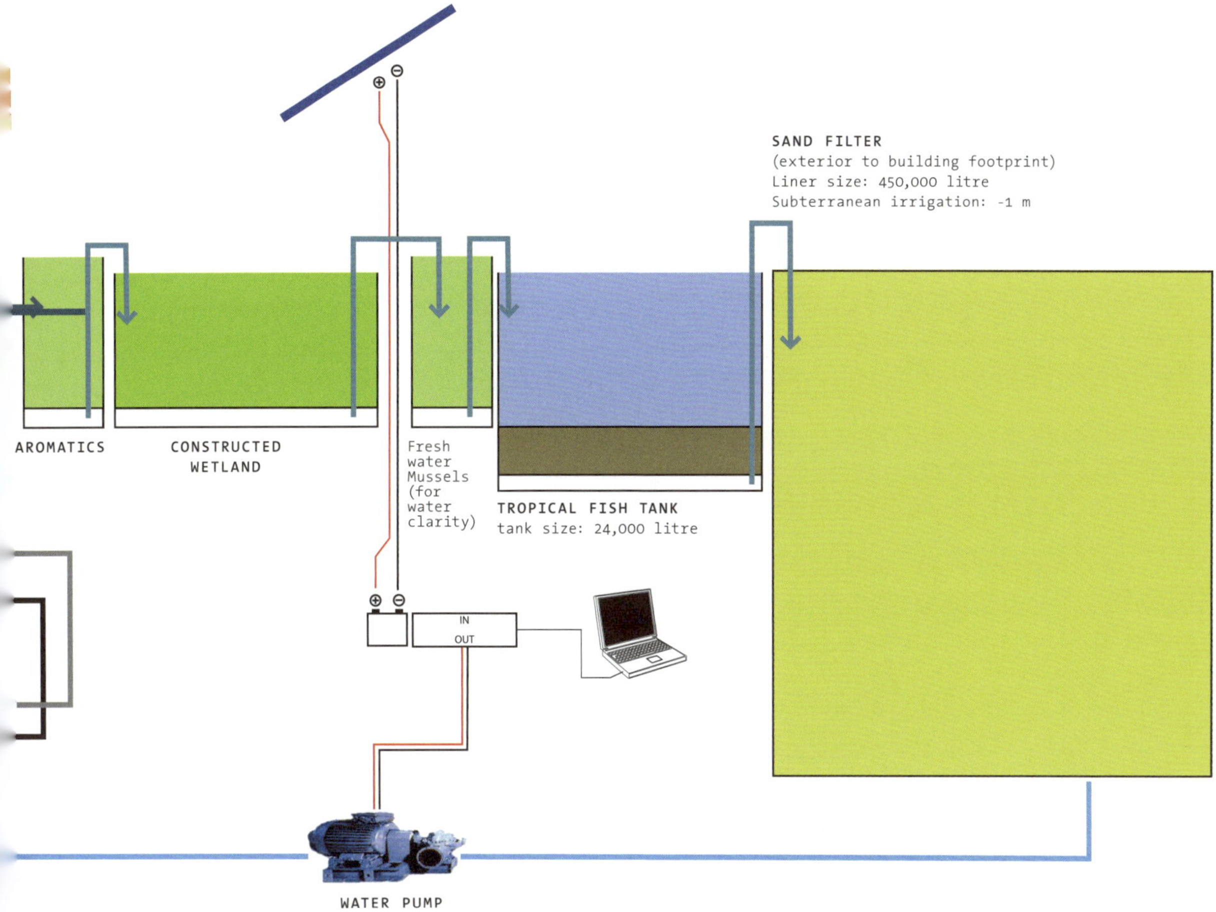

/06
Diagram of water filtration loop

The newly ingested yard/lawn becomes a lush, living filtration system, cleaning grey and black water for reuse, while offering an Eden-like garden for its inhabitants. Eden is sustained by the toilet.

View from basement into treatment tanks below interior yard. The sublime is a by-product of the sustainable machinery.

rgb garden

The garden consists of 850, 15cm x 15cm x 15cm algae photo bioreactors. Half-machine, half-alive, these are plant cyborgs / **Studio NMinusOne with Nashid Nabian**

The reactors sequester carbon dioxide with an efficiency far exceeding that of a tree. Each reactor opportunistically gathers energy using ground, wind or solar harvesting technologies.

During the day, energy is captured and stored from these various sources depending on weather conditions. During the night, it is released as artificial light, extending the algae's growth cycle past the day's sunlight hours. The algae's growth is artificially accelerated/augmented beyond its natural rate. The pattern of illumination reflects the energy source it was powered by, Green = Earth (moisture), Blue = Wind and Red = Sun (Fire). The garden becomes a map, a visual recording of the day's weather conditions. During the course of the International Garden Festival 'Jardins De Métis', the garden will be in constant flux; the algae will visibly grow at different rates daily, changing in density and depth of colour to offer a new experience every day.

/01
Plan Diagram Day Condition – Algae growth depends on environmental factors: wind, sun and rain. The plan becomes a record of past weather – long term.

/02
Plan Diagram Night Condition – At night the LEDs turn on and use the electrical charge gleaned from solar, wind and rain energy. The light configuration becomes a display of daytime weather patterns – short term.

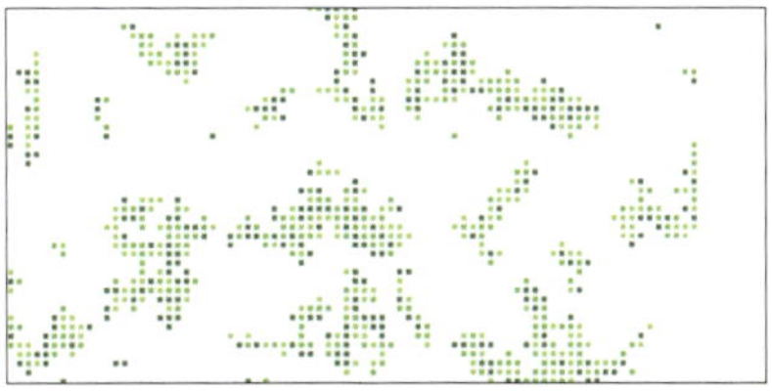

/01

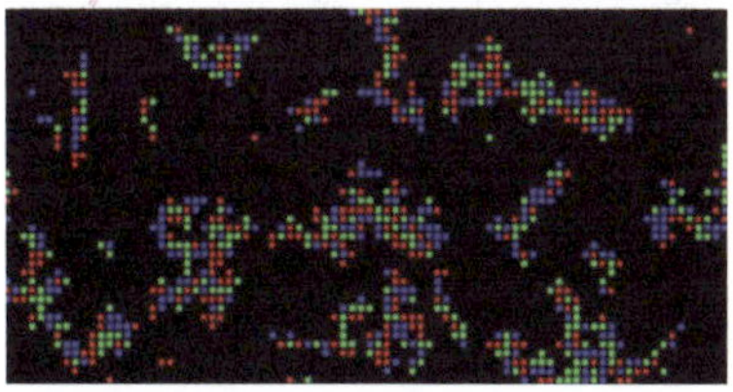

/02

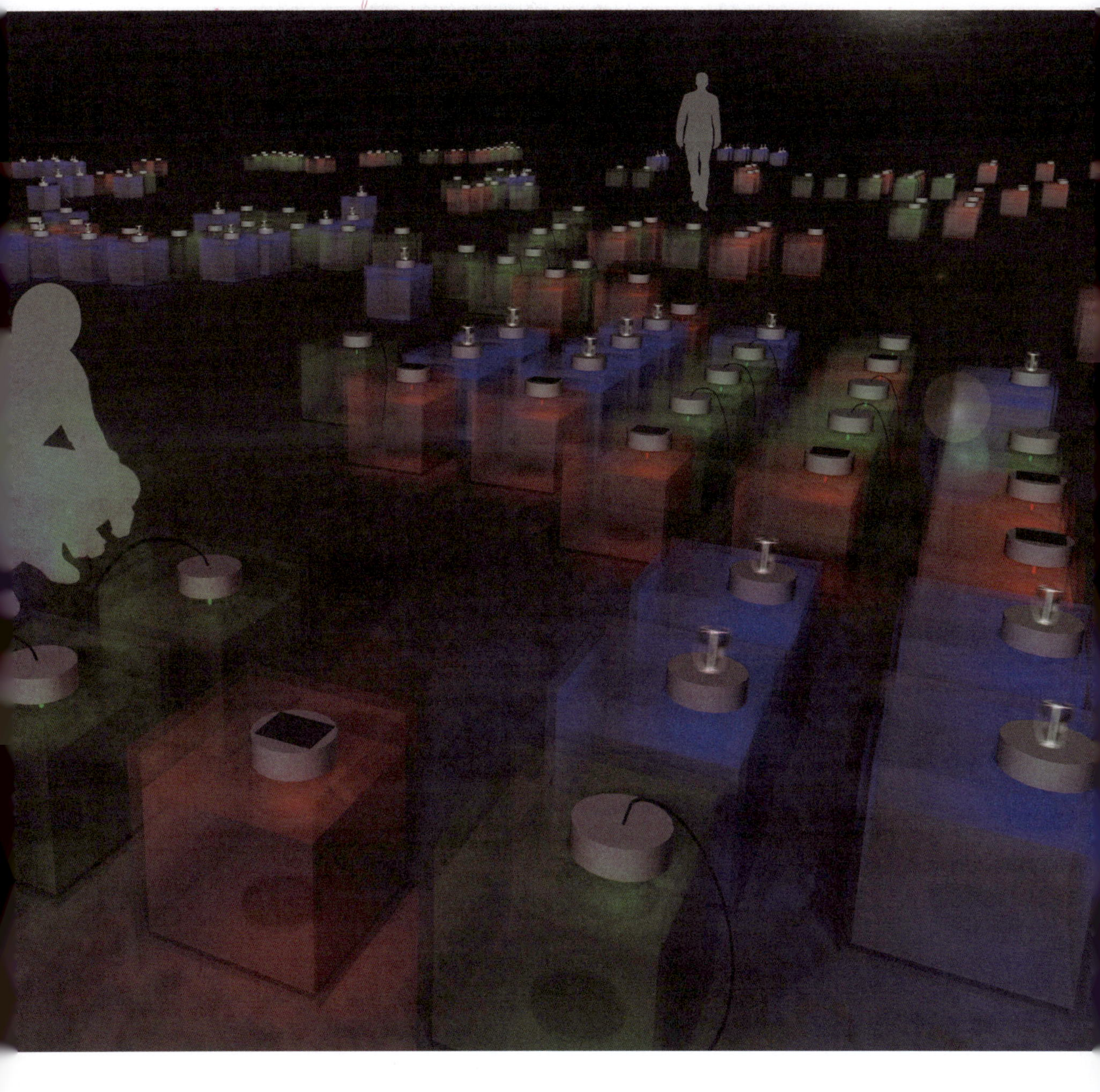

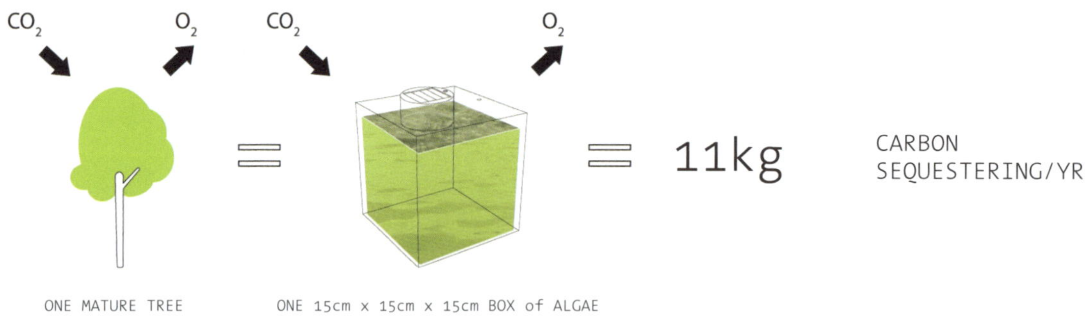
CO2
O2
CO2
O2
=
=
11kg
CARBON SEQUESTERING/YR
ONE MATURE TREE
ONE 15cm x 15cm x 15cm BOX of ALGAE

/03

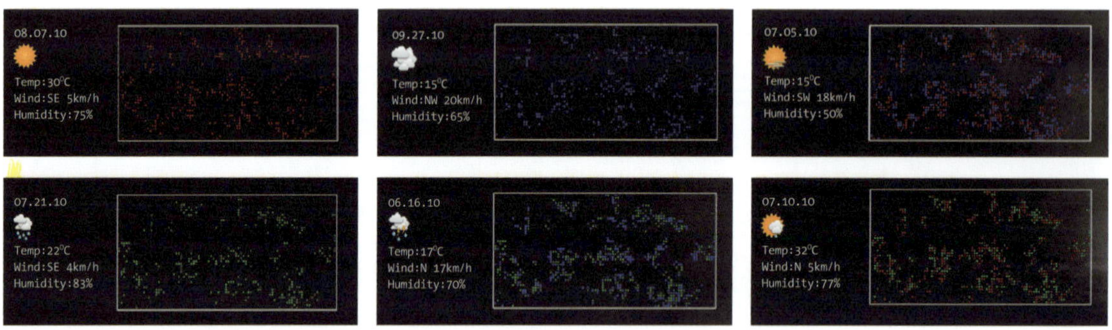
08.07.10
Temp:30°C
Wind:SE 5km/h
Humidity:75%
09.27.10
Temp:15°C
Wind:NW 20km/h
Humidity:65%
07.05.10
Temp:15°C
Wind:SW 18km/h
Humidity:50%
07.21.10
Temp:22°C
Wind:SE 4km/h
Humidity:83%
06.16.10
Temp:17°C
Wind:N 17km/h
Humidity:70%
07.10.10
Temp:32°C
Wind:N 5km/h
Humidity:77%

/04

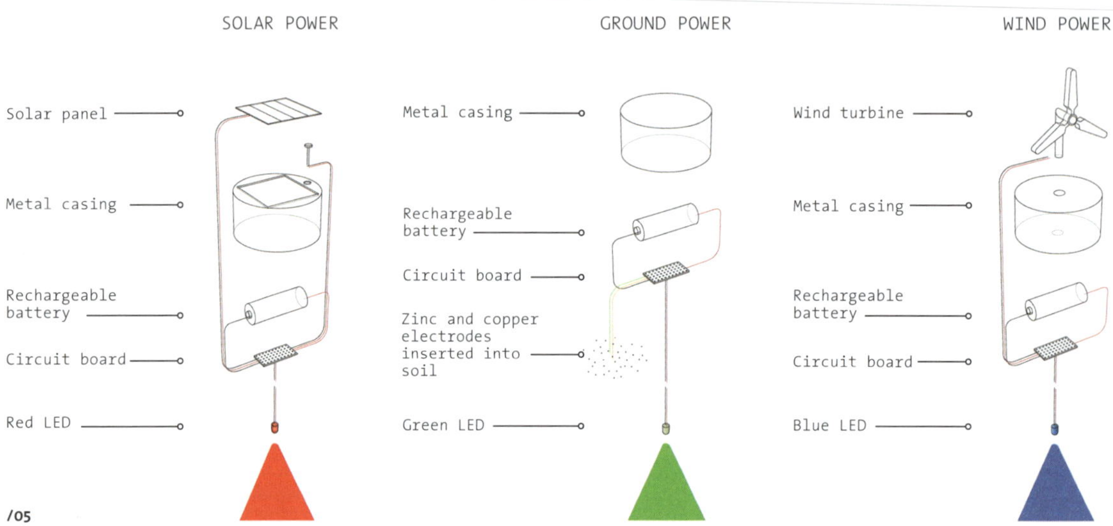
SOLAR POWER
Solar panel
Metal casing
Rechargeable battery
Circuit board
Red LED
GROUND POWER
Metal casing
Rechargeable battery
Circuit board
Zinc and copper electrodes inserted into soil
Green LED
WIND POWER
Wind turbine
Metal casing
Rechargeable battery
Circuit board
Blue LED

/05

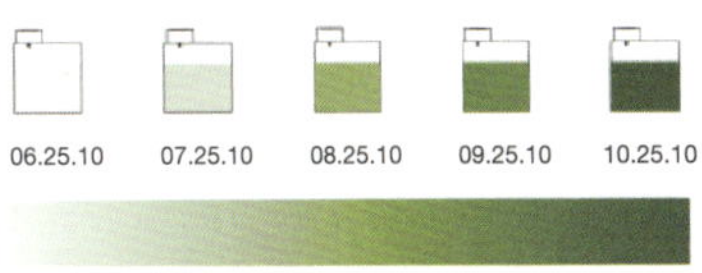

/06

/03
One box of algae growing at optimal levels can sequester as much CO2 as one mature tree.

/04
The garden becomes a map, a visual recording of the day's weather.

/05
Three different energy sources are harvested. Ground, wInd and solar: earth, wind and fire.

/06
Algae growth rates will vary depending on the weather. Over an extended period of time the predominant source of energy will become visible though the densest algae growth.

immersive spaces

Architecture's traditional capacity to shape and condition space is expanded through technologically enhanced physical components. Spaces are created not only with material elements such as walls and roofs, but also by means of non-substantial effects. The precise technological manipulation of climate, atmosphere, and ambience generates sensorial, all-enveloping environments that redefine our experience of the world. These projects explore molecular, physical and psychological elements to generate highly modulated atmospheres. They are finely tuned to respond to the user and resonate with the dynamic fluctuations of the environment, paving the way for a more socially and ecologically engaged architecture.

sky house

The Sky House achieves one of the oldest ambitions of architecture… To suspend the body in space
/ **Studio NMinusOne**

The architectural components used to achieve this are also the essential ingredients for a low-carbon footprint house in arid regions.

Sustainability in the service of pleasure. The house is half house, half pool. The interior temperature of the house is regulated by the pool water. The exterior wall of the pool forms the interior walls of the house. The wall works as a heat exchange device. To maximise exchange it is shaped as a series of stacked, continuous, cooling ribs, assuming a form for maximum surface to air exchange. The filtration system of the pool runs the water through a tank of Phase Change Materials (PCM) encapsulated in 7.60 cm diameter plastic capsules. During the day the PCM absorb the solar energy embedded in the water from the sun. At night the PCM release this energy back into the water and subsequently back into the house.

/01

/01
The roof is covered by 2.50cm of water, making the pool, hot tub and roof into a single reflective surface. The sky and its reflection in the water merge seamlessly. People walking on the roof deck appear to be floating in space.

/02

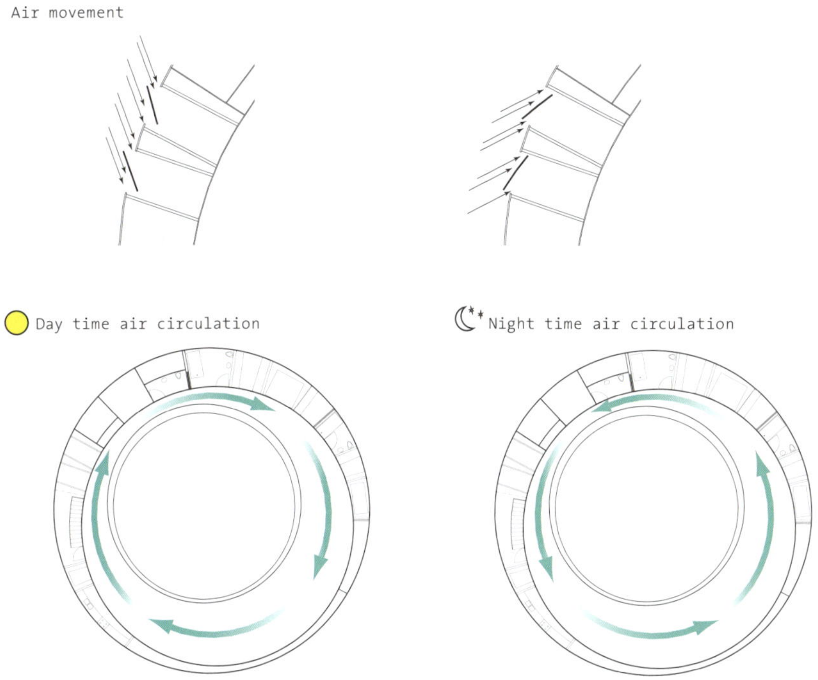

/03

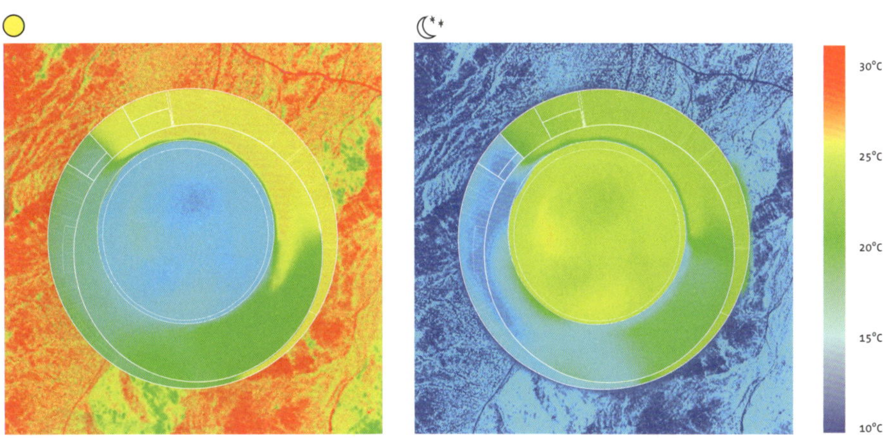

/04

/02
The roof surface is painted black to increase the reflectivity of the water surface. A combination of mist makers and ultrasonic atomisers eliminate the horizon line.

/03
During the day, the temperature difference between the hot exterior climate and the cool pool water generates clockwise thermo-dynamic air circulation within the living space. At night the temperature difference between the cold exterior and the warm pool water generates a counter-clockwise air circulation pattern. The air circulation keeps the space warm at night and cool during the day.

/04
Circulation diagram of pool water and phase change materials.

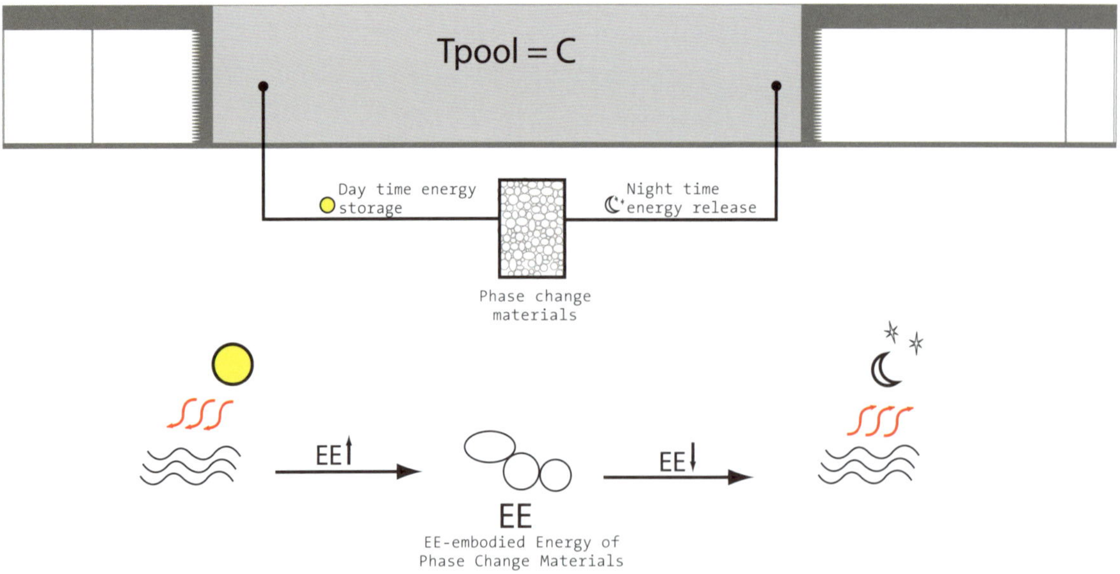

/05

/05
The ribs around the central pool act as heat exchangers, regulating the temperature of the house.

/06
Bottom of the hot tub as seen from the living room.

/06

all seeing house

Building bricks that mimic the compound eye, embedded with sensor technology to create an engineered transparency / **Justin Cheng***

In the All Seeing House, individual brick units are equipped with an array of the latest sensor technology, capable of detecting gamma radiation, ultraviolet radiation, heat flux, moisture levels, and carbon dioxide levels.

The All Seeing Bricks mimic the compound eyes of insects, each facing in a slightly different direction, thereby projecting the visual range of a dome onto a box. The bricks transmit the information to an interior surface of correlated RGB LEDs, producing an engineered transparency to the outside world.

The house is conceptualised as a scheme divided into three spaces: the outside world, an interior void that produces the immersive, heightened experience of the outside world, and the space between the two, where the residential programmatic requirements are packed tightly alongside – if not relegated to the status of – the mechanical infrastructure of the house. Void tubes that connect the sensor skin to the cosmic display not only allow circulation and entry of natural light, but also express the omnidirectional vision of the house.

/02

/01

/01
Next page. The main living space is bathed in a continuously changing kaleidoscope of colour which is coded according to the natural phenomena that occur outside.

/02
The sectional model reveals the interior living chamber.

*Guided by Christos Marcopoulos and Carol Moukheiber

/03

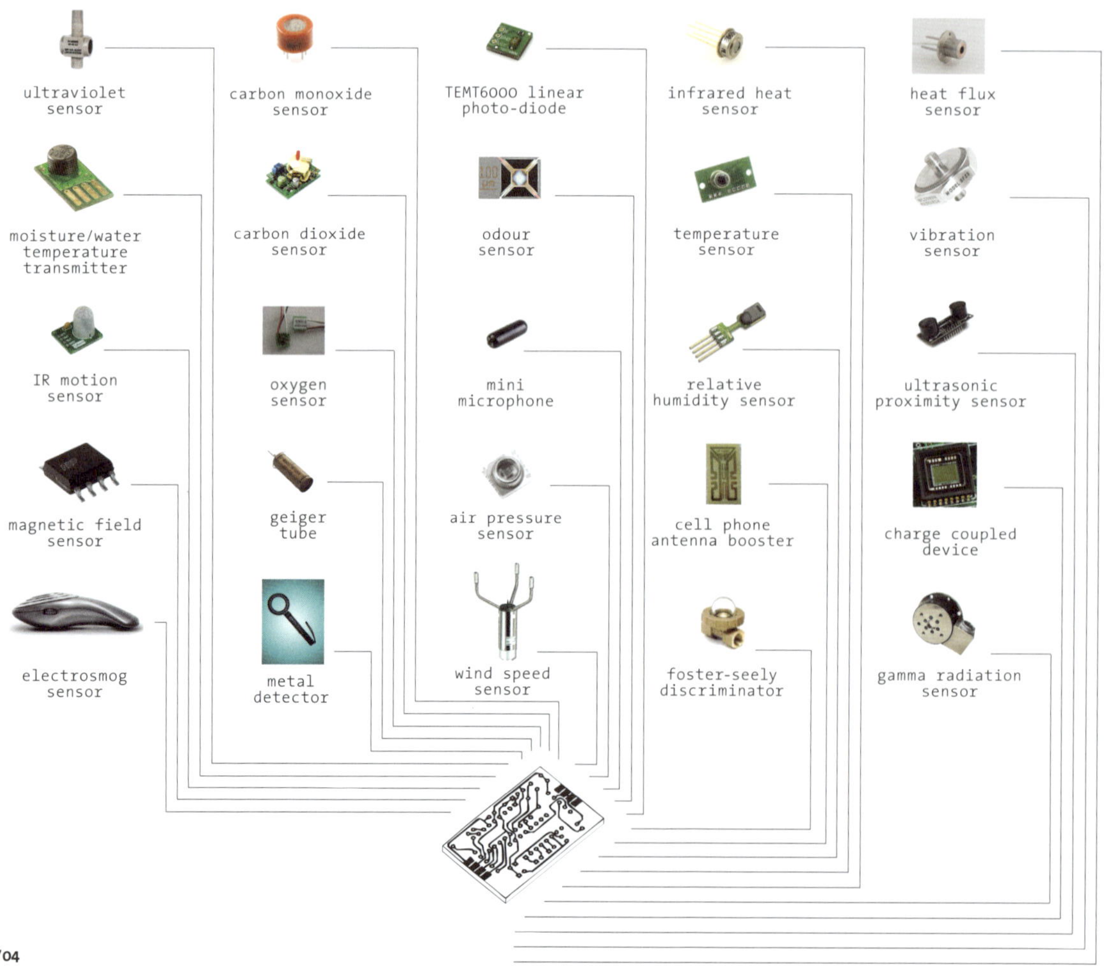
ultraviolet sensor
carbon monoxide sensor
TEMT6000 linear photo-diode
infrared heat sensor
heat flux sensor
moisture/water temperature transmitter
carbon dioxide sensor
odour sensor
temperature sensor
vibration sensor
IR motion sensor
oxygen sensor
mini microphone
relative humidity sensor
ultrasonic proximity sensor
magnetic field sensor
geiger tube
air pressure sensor
cell phone antenna booster
charge coupled device
electrosmog sensor
metal detector
wind speed sensor
foster-seely discriminator
gamma radiation sensor

/04

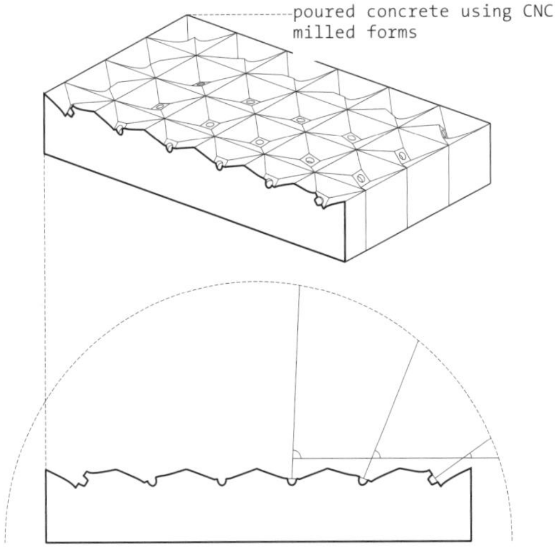

/05

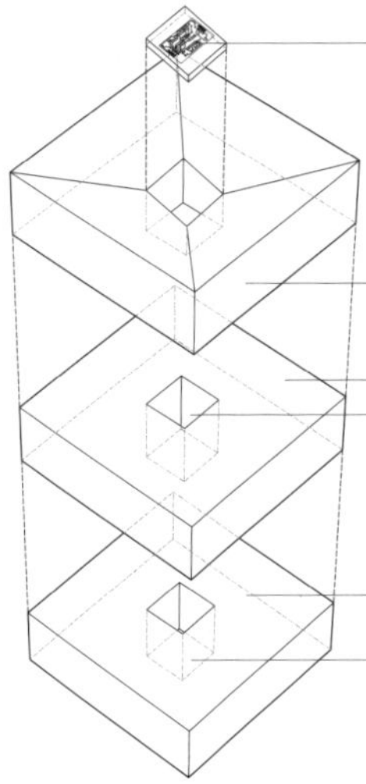

/06

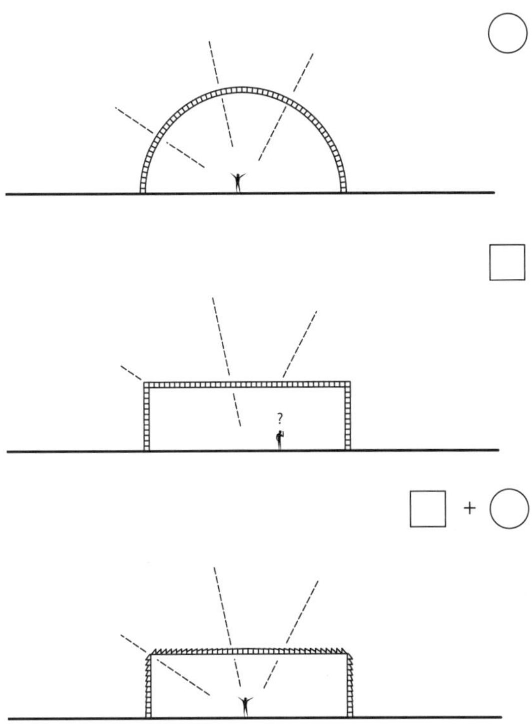

/07

/03
Sections exposing the relationship between the inner chamber and the outer walls.

/04
A multiplicity of sensors are incorporated into the 'brick'.

/05
Each brick is angled to respond to the sun's path. A flat roof effectively mimics a domed surface.

/06
Brick assembly explained.

/07
The capacity for a 360 degree panoramic vision afforded by the dome is transferred to each brick. By arraying the angle of each brick, the box acquires the all-seeing performance of the dome, while maintaining the flexibility of a rectilinear space.

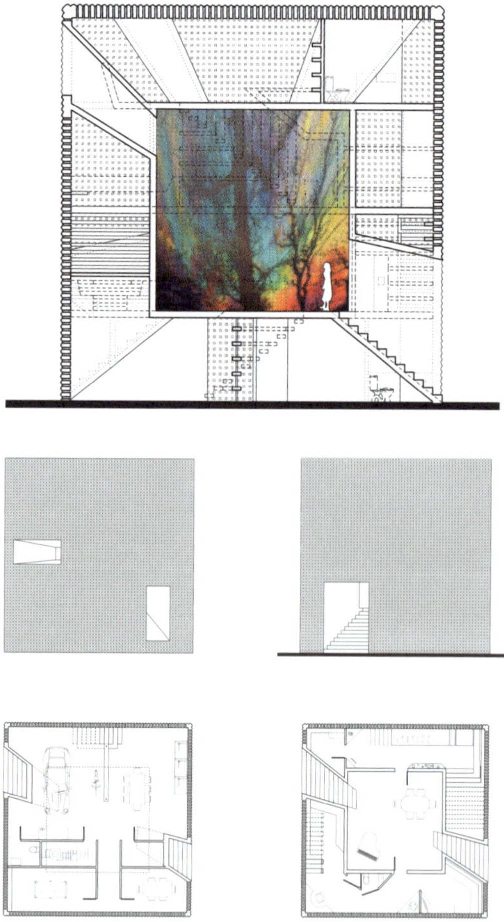

/07

/08

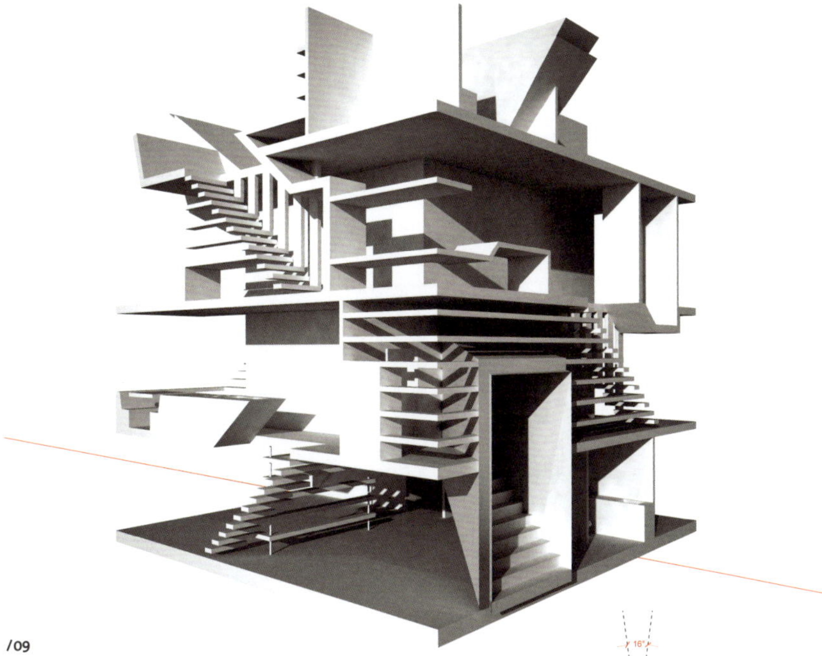

/09

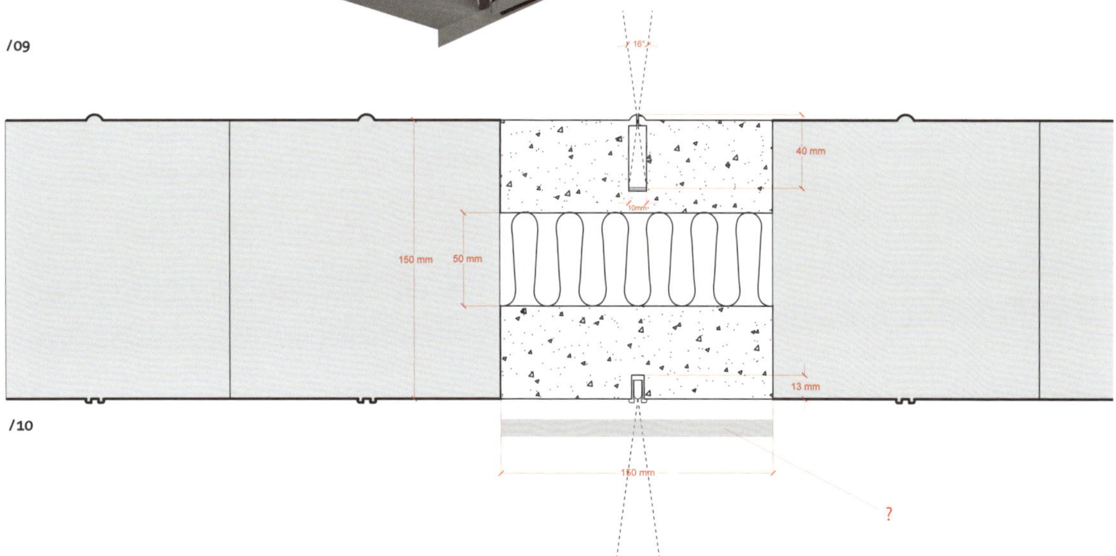

/10

/07
Elevations, plans and section.

/08
Elevation.

/09
Elevations, plans and section.

/10
Wall section.

tunable sound cloud

An acoustic liner designed to enhance one's auditory experience / **Mani Mani***

The Tunable Sound Cloud (TSC) is an acoustic liner designed to enhance auditory experience within a generic space. Comprised of triangulated panels that dynamically change shape to adjust to various acoustic needs, it is capable of simulating highly specific auditory spatial configurations.

Up until the early 20th century, there was an intimate relationship between an architectural space and the music generated within it (chamber music in palace chambers, Baroque music in opera houses, etc.). Contemporary performance halls typically achieve flexibility in acoustic effects through the strategic placement of acoustic panels.

Actuated with a matrix of shape memory alloys and stepper motors – controlled with Arduino micro-controllers – the Tunable Sound Cloud responds in real-time to the acoustic needs of its environment. It can thus simulate a range of auditory spatial configurations. Auditory data is streamed through microphones and analysed through MAX/MSP software which adjusts the skin surface for sound reflection/absorption in different scenarios. The architectural surface becomes yet another instrument for the deployment of an acoustic effect.

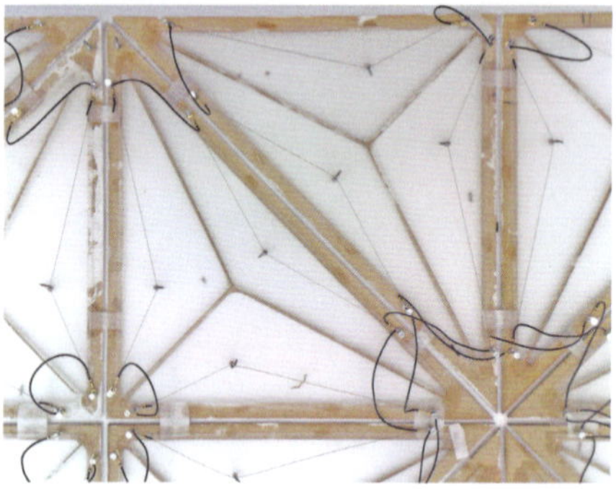

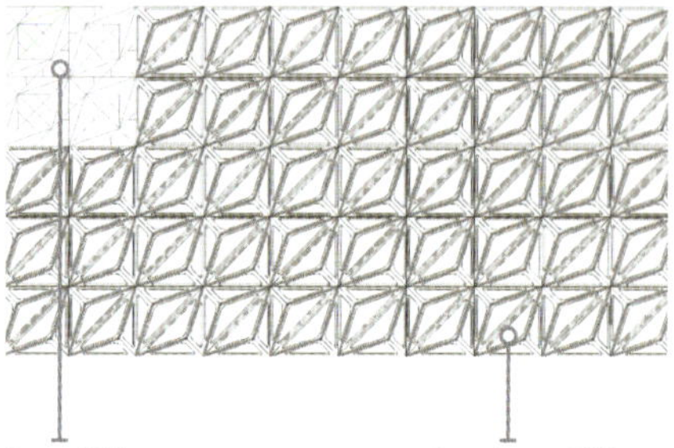

/01

/01
The prototype materials consist of basswood, mylar, cables, and muscle wire or shape memory alloy.

/02
The Cloud application can be manually adjusted for customised spatial moments, or programmed to allow the software to autonomously calibrate the system through a series of feedback loops. TSC's modular design allows for large scale expansions and site specific installations.

*Supervised by Christos Marcopoulos and Carol Moukheiber

/02

breeze-thru wall

These permeable wall modules are an alternative to the sealed and insulated house / **Arthur Tseng***

The Breeze-Thru Wall explores the potential for ventilation and heat recovery to be integrated within a performative building module.

The wall consists of modules acting as air intake units which lock together for structural integrity. The modules are part of an overall envelope system of exchanging and recovering indoor heat. Each module is a funnel designed to capture incoming air efficiently. The air is then held within copper tubing for heat exchange before entering the interior wall. The cycle of hot and cool air from exhaust and intake provides constant ventilation and recovery.

The heat recovery ventilator unit, designed at the scale of the house, is run in response to exterior wind conditions. The module, along with the heat recovery cycle, can be reversed as the prevailing wind changes; the two exterior funnel-shaped faces of the module act as both intake and exhaust openings. The heat recovery cycle uses forced air, which runs on a loop around the envelope of the house, carrying the necessary heat and cold to be exchanged with each module.

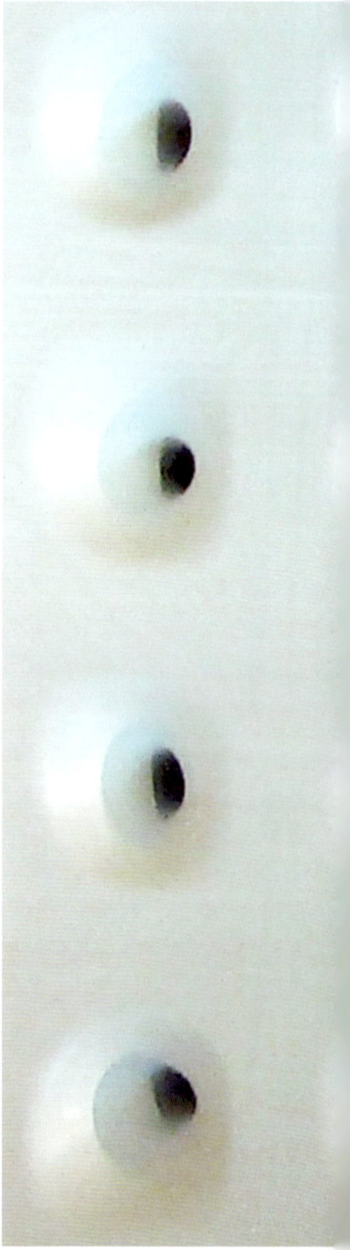

/01

/01
Wall elevation, interlocking modules.

*Guided by Christos Marcopoulos and Carol Moukheiber

/02

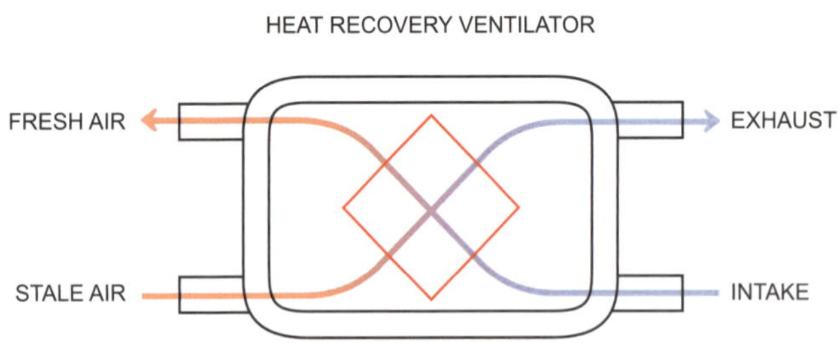

/03

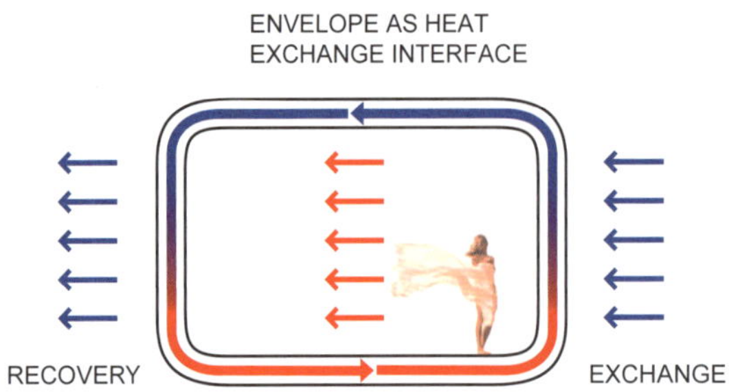

/04

/02
Incoming cool air turns into a warm breeze as it passes through the building envelope.

/03
The traditional set-up of the Heat Recovery Ventilation Unit. Warm, stale exhaust air heats up incoming cooler fresh air, thus making fresh air ventilation more efficient.

/04
Exchange Cycle. Incoming cool air blows through the wall. In the process, it is heated by warm air housed inside the wall // Recovery Cycle. Interior warm air exits through the wall. In the process, it heats up the air housed inside the wall.

/05
Anatomy of a module.

/06
Prototype fabrication process using thermoplastics.

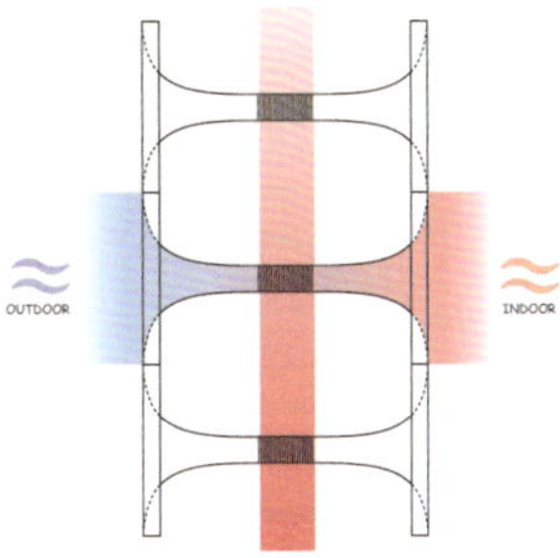

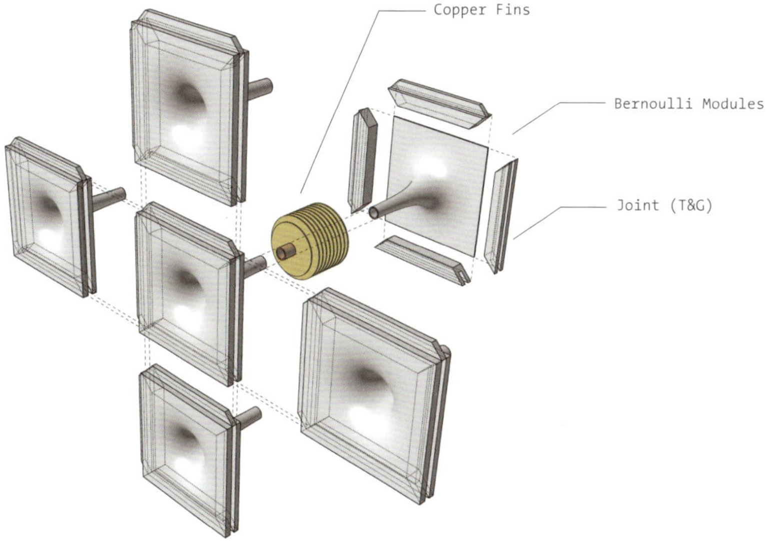

/05

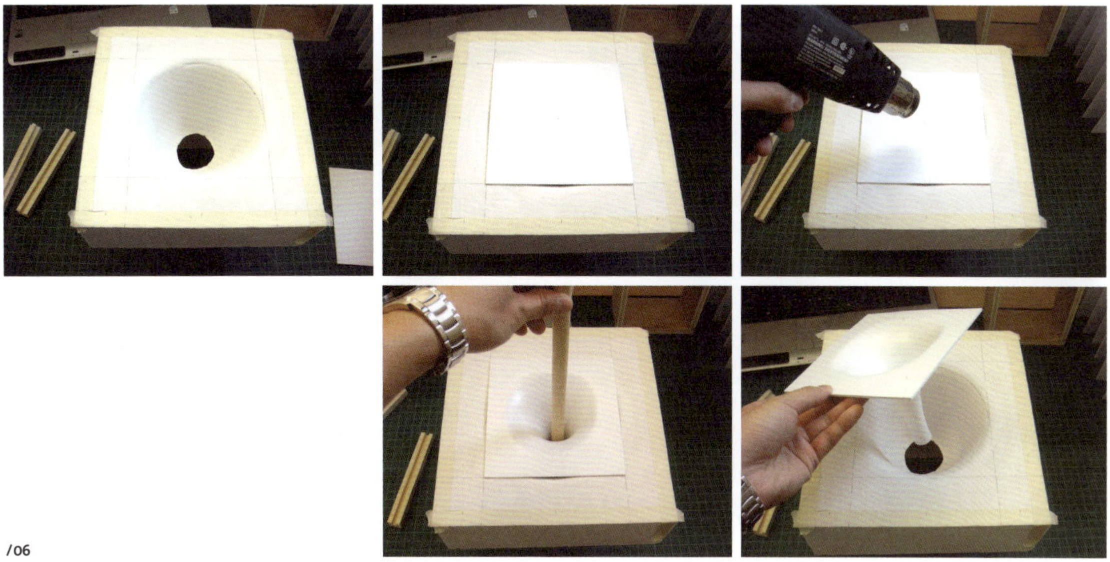

/06

play on the fly

Releasing the body from its routine composure, daily 'Body Play' opportunities are created in the public realm / **Lailee Soleimani***

Play on the Fly is an interactive, cybernetic, ambient system that offers opportunities for 'play' in the public realm. Building on technological advances in computationally administered gesture recognition, the system recognises various playful body gestures and responds to them by activating an experiential sequence.

Influenced by Piaget's play theory, Play on the Fly investigates the role of digitally augmented architectures in creating daily body play opportunities for adults where 'play' becomes part of their busy routine life and a 'purposeless purpose' is added to the act of playing. The selected site is Toronto's PATH system, a subterranean labyrinth that connects downtown Toronto. The design introduces a network of interactive ambient playgrounds in the corridors of the PATH. The playgrounds contain adventures hidden within the existing programmes of the urban network. They incorporate a series of interactive games where a movement in a certain direction activates a screen on which the player can navigate through the 3D model of the site, or get contextual information. Interactive screens are hidden inside the structure of the PATH's public corridors and only become activated when a person starts playing. Hence, anyone can choose to be a player or an audience member.

The proposed ambient game system relies heavily on innovations in computer vision, eliminating the need for interacting with a gaming console by incorporating algorithms that interpret the body's gestures as user input. Once a particular gesture is recognised by the system, it activates a certain play scenario that is conveyed to the user via the embedded digital displays.

The proposal is a series of activities that are inspired by conventional playground equipment. Each activity is an interactive game where a body movement in a certain direction activates a screen on which the player can navigate through the 3D model of the site or get contextual information by moving his body. The interactive screens are hidden inside the public corridors of PATH system as part of their structure: walls, floors, and ceilings, and only become activated when a person starts playing.

The playgrounds are hidden adventures within the existing program of the urban network, and the play activities of each individual become departures from the everyday routine of navigating the corridors to and from work, as anyone can choose to be a player, or an audience member along the way.

/01
A scene from the underground PATH network. As the user twirls in place, a worm hole appears on the ceiling becoming larger as the player twirls. The scene on the ceiling allows one a glimpse into the exterior, digitally reconnecting the artificially lit interior with the outside.

*Supervised by Rodolphe el-Khoury

/01

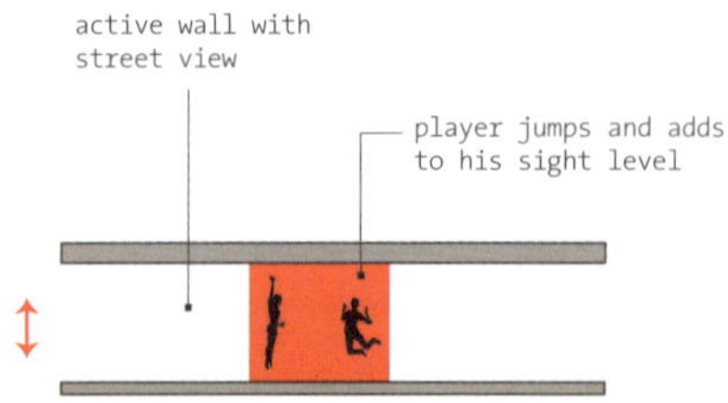
active wall with
street view
player jumps and adds
to his sight level

/02

THE SKY-TWIRL

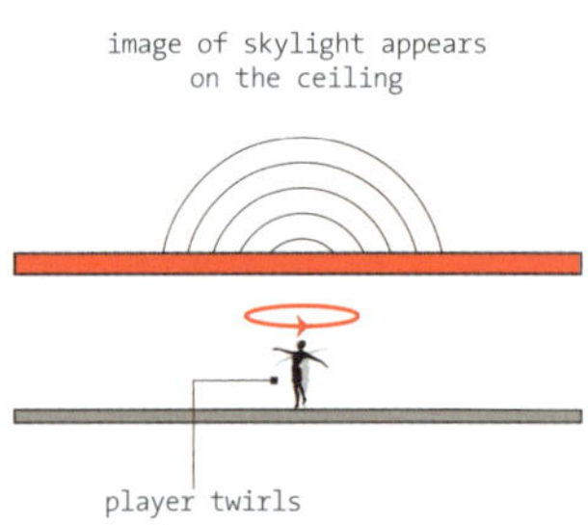
image of skylight appears
on the ceiling
player twirls

/03

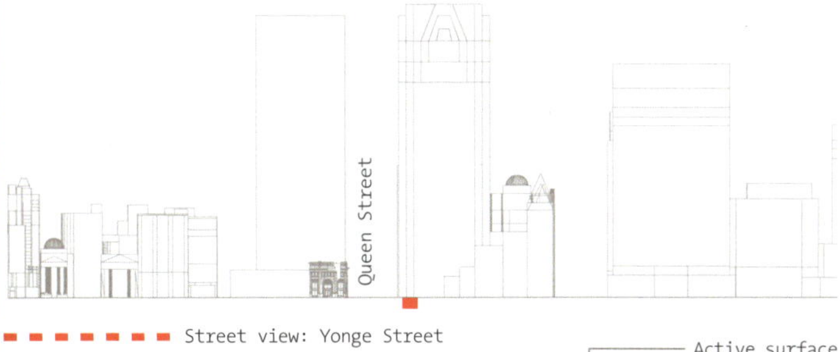

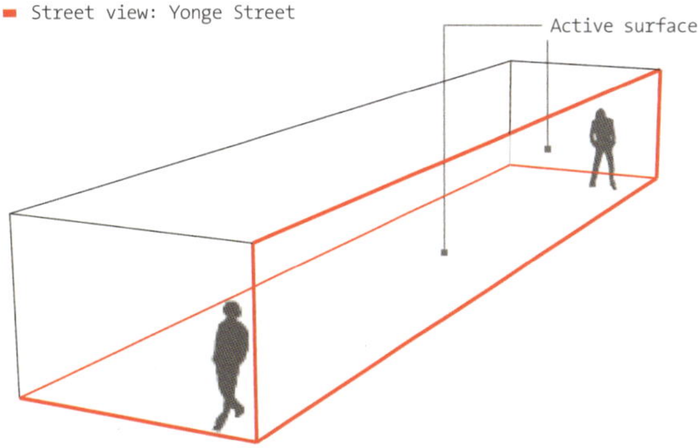

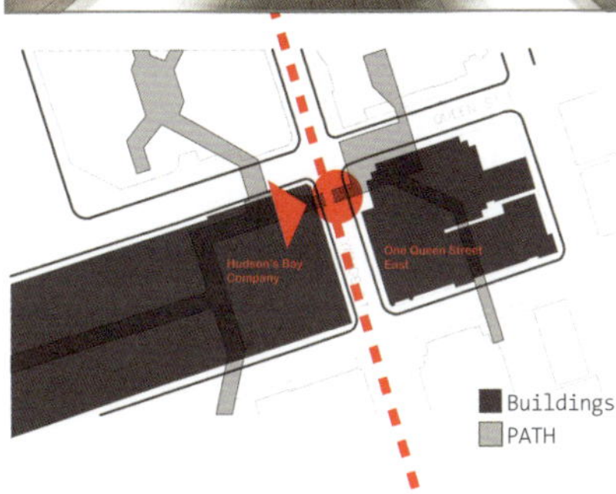

Dimensions (wxlxh): 3.60 x 19.80 x 2.30m
Space Analysis: Equal movement in both directions. Low Ceiling.
Direct Access: One Queen Street East. Bay Company.
Close Major Destinations: Eaton Center, Queen Subway, Bay Company.

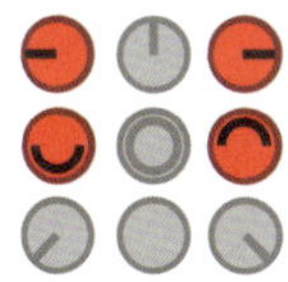

/04

/02
Scenario one. When a player jumps in front of a wall, a screen is activated. The screen shows the player. As the jumping continues, the screen shows a higher view, as if the player is adding to her height with each jump.

/03
Scenario two. The player begins by twirling. An image of a skylight appears on the ceiling above in a small circle. As the player continues to twirl, the circle becomes larger.

/04
Diagram showing location and purpose of active display surfaces.

/05

/05
Three different scenarios are explored within the PATH system, a subterranean labyrinth which connects downtown Toronto.

invisible house

In the Invisible House, lighting is the primary shaper of one's spatial experience / **Studio NMinusOne**

The notion of the invisible is explored in a couple of different ways: on the one hand the extreme familiarity and homogeneity that renders things invisible and on the other, the literal white out.

On the interior, architecture is systematically removed from one's visual perception, only the objects which escape any form of permanence will appear. During the night, the living room is a black void, and during the day it becomes a white void. At night, only the essential is illuminated. Objects are detected by infrared beams, and illuminated according to their presence and location. The floor and ceiling are composed of 3.50 cm x 3.50 cm corresponding tiles. The ceiling tile contains a grid of 16 LED lights with infrared sensors.

The floor tile has 16 infrared receivers which correspond to the ceiling tile sensors, creating a field of invisible vertical infrared beams within the space. When an object interrupts those beams, the corresponding ceiling LED narrow spotlights turn on.

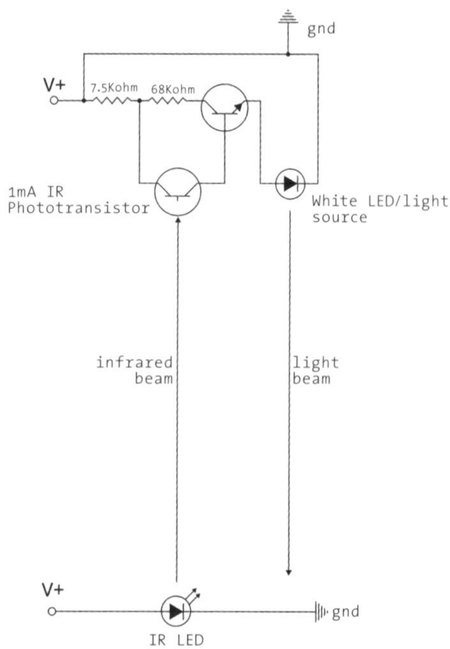

/01

/01
Circuit schematic.

/02
Urban Invisible House. Rodeo Drive, L.A., 2007.

/03
Ceiling and floor tiles dynamically illuminate a person walking.

/02

/03

TIME

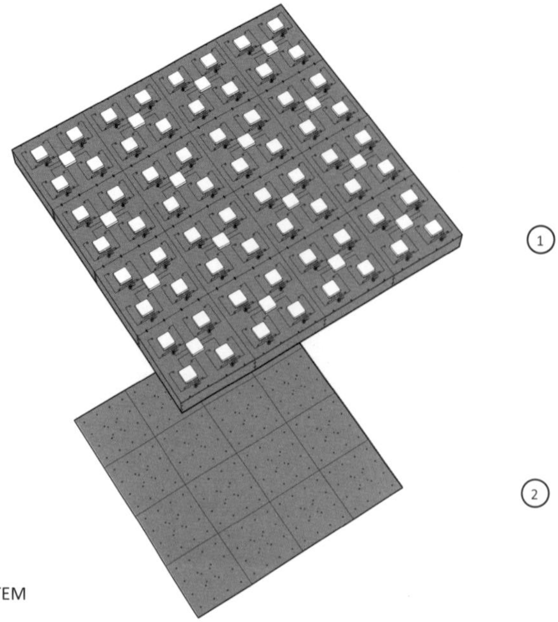

LIGHTING SYSTEM

(1) ceiling panel
(2) floor panel

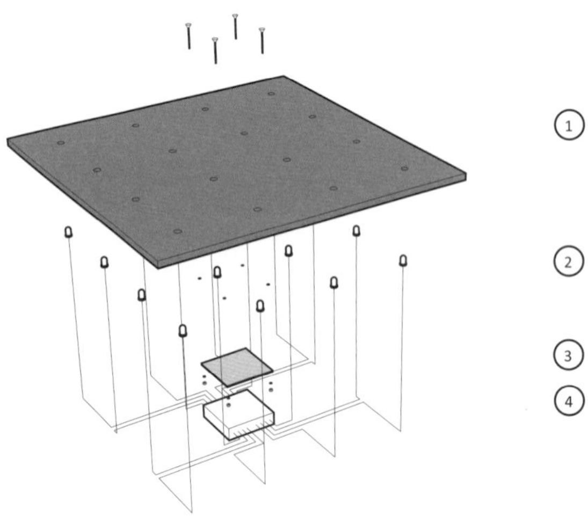

FLOOR PANEL

(1) frame
(2) IR emitters
(3) circuit substrate
(4) power distribution circuit

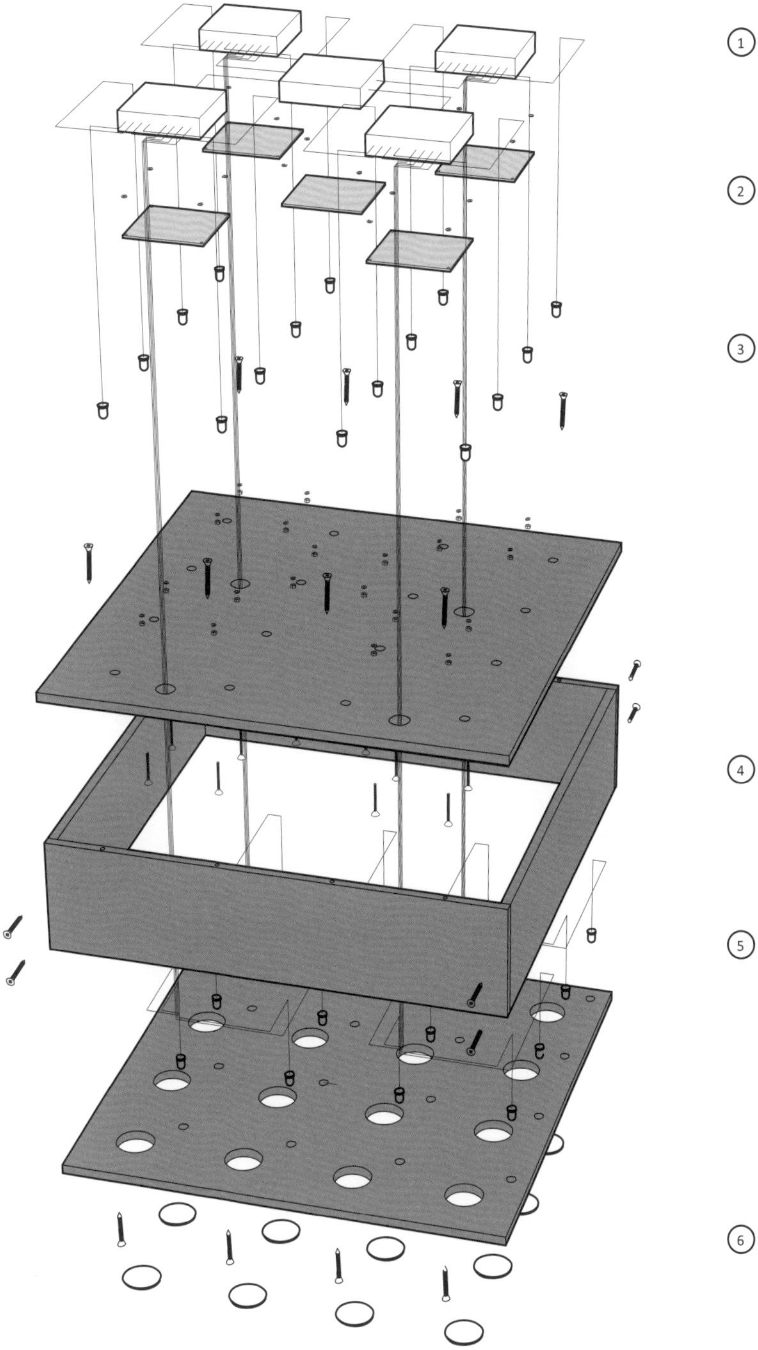

CEILING PANEL

1. comparator and power circuits
2. circuit substrate
3. high intensity white LEDs
4. frame
5. IR detectors
6. lenses

/04

/05

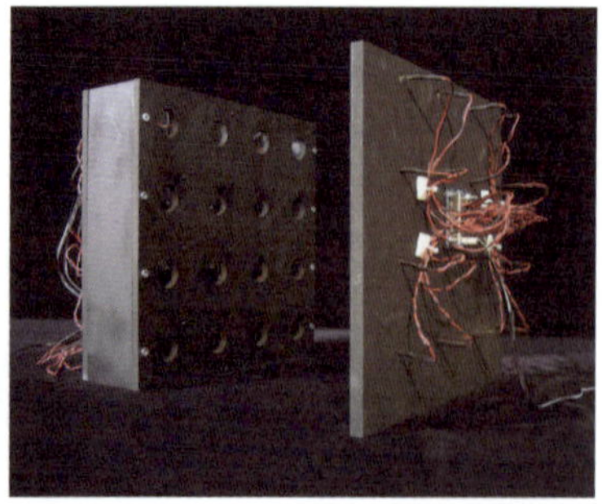

/06

/07

/04
The first prototype of the ceiling tile. The materials used included foamcore, cardboard tubing, and an array of lenses.

/05
The first prototype of the floor tile. Infrared sensors are aligned with the ceiling tile (all lights are on).

/06
The second prototype of the floor and ceiling tiles. A rigid wood frame replaced the foamcore.

/07
Through a form of suburban camouflage, the exterior of the house appears as being invisible too.

/08
Section A-A and B-B. The interior spaces of a typical suburban spec house are reshaped to create rounded corners. The soft edges between wall, floor and ceilings turn the interior into a virtual void.

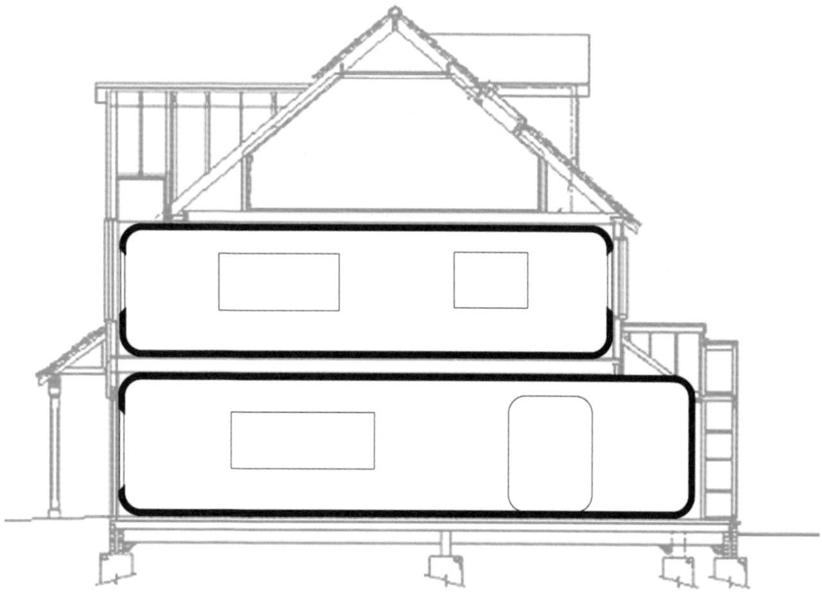

/08

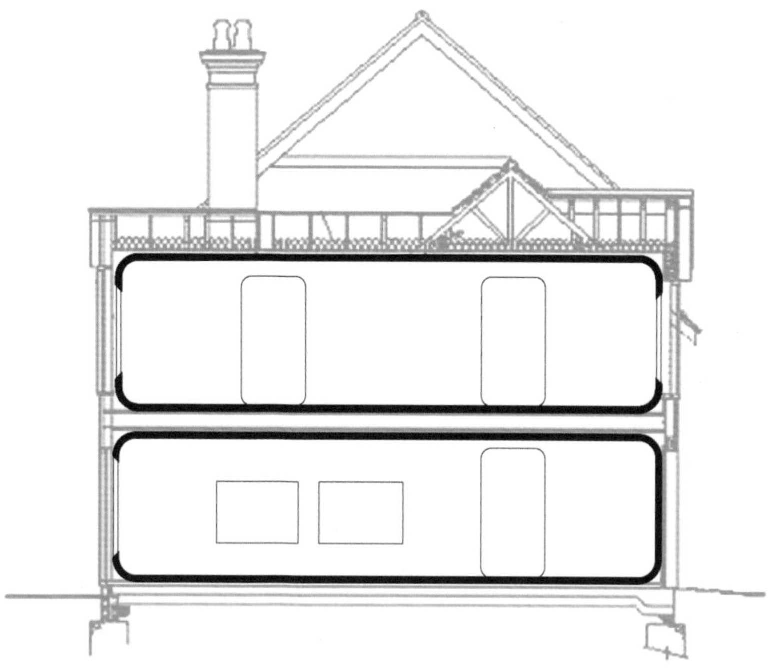

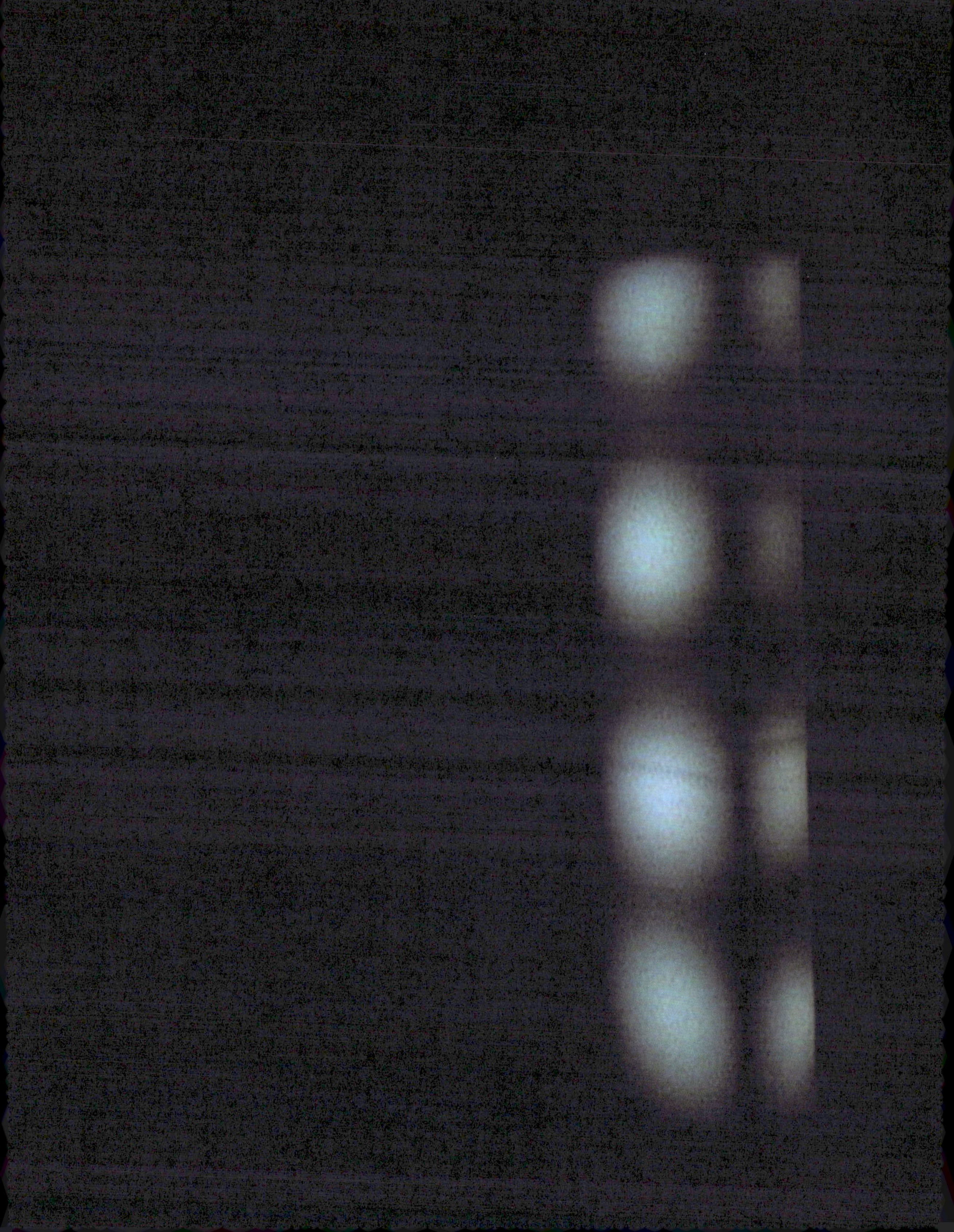

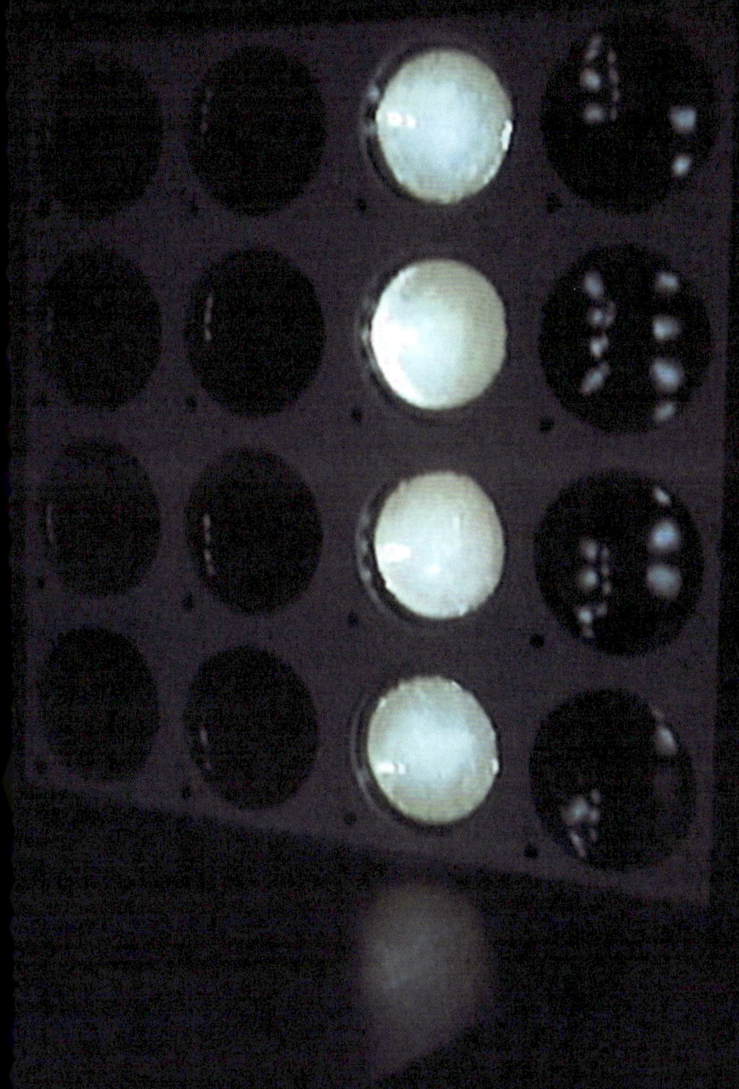

farnsworth wall

The Farnsworth Wall is an architectural wall module that harvests solar energy to illuminate the interior / **Studio NMinusOne with Mani Mani**

Through the use of solar panels on one side and LED lights on the other, an opaque wall becomes digitally transparent.

The Farnsworth Wall takes its cue from Mies van der Rohe's Farnsworth House. The famous glass house made its inhabitants uneasy with its complete transparency to the outside. Perimeter curtains were hung to provide opacity and psychological comfort. With the Farnsworth Wall, the outside becomes a dynamic digital wallpaper allowing one to see the outside without the outside seeing in.

The wall consists of a modular panelling system built as a Structural Insulated Panel (SIP) in a standard 1.2 m by 2.40 m size. This module can be cut and applied to virtually any existing or new construction. A layer of solar panels on the exterior of the module harvests and stores radiant energy, while an embedded system of low-energy LED lights on the other side illuminates the interior space.

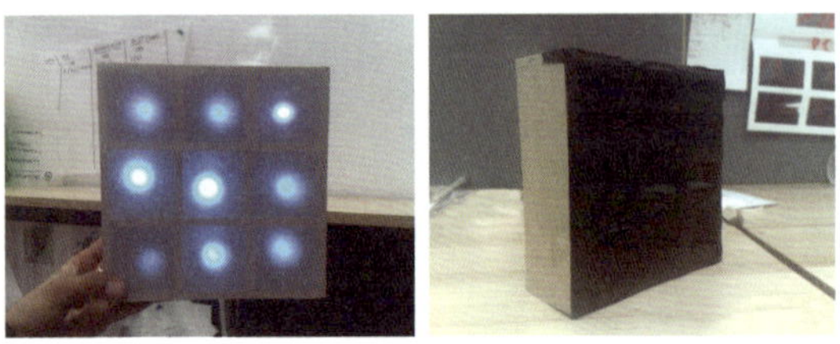

/02

/01

/01
Bodies or shadows disrupt the passage of light on the exterior surface, causing the LED lights to dim in the corresponding areas on the interior. This gives the effect of one-way transparency.

/02
A 1:1 prototype of a portion of the Farnsworth Wall module: interior and exterior.

farnsworth curtain

The curtain forms an enclosure yet allows the outside to register itself on its interior surface / **Studio NMinusOne with Rodolphe el-Khoury**

The Farnsworth Curtain is a soft version of the Farnsworth Wall project. The curtain carves out a curvilinear space which allows for both open and more intimate spaces.

The curtain forms an enclosure yet allows the outside to register itself on its interior surface. Through a grid of infrared sensors and LEDs, objects and bodies on the exterior of the curtain appear as shadows on the curtain's interior surface – creating transparency through electronic mediation. The space within is differentiated further by dampening all sound. The enclosed space then becomes a sensory refuge from the bustle of the main floor, while still maintaining contact.

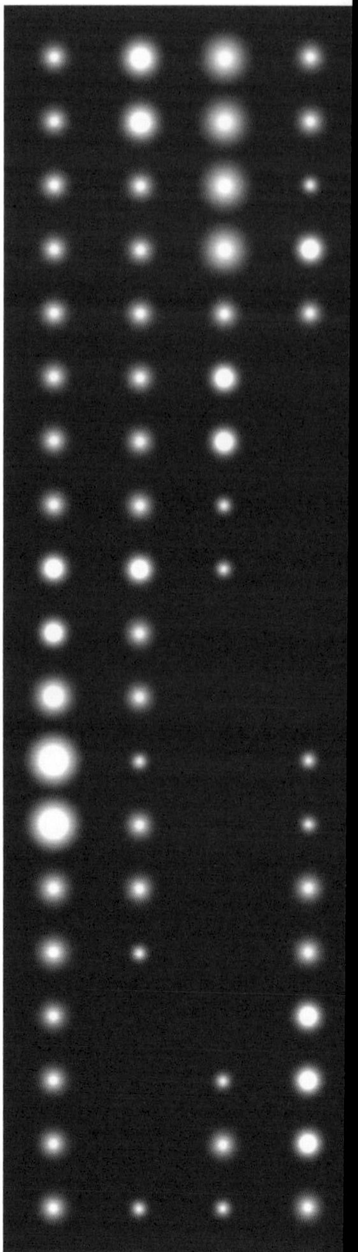

/01

/01
Shadows will seemingly float by, the busier the outside the more hypnotic the experience.

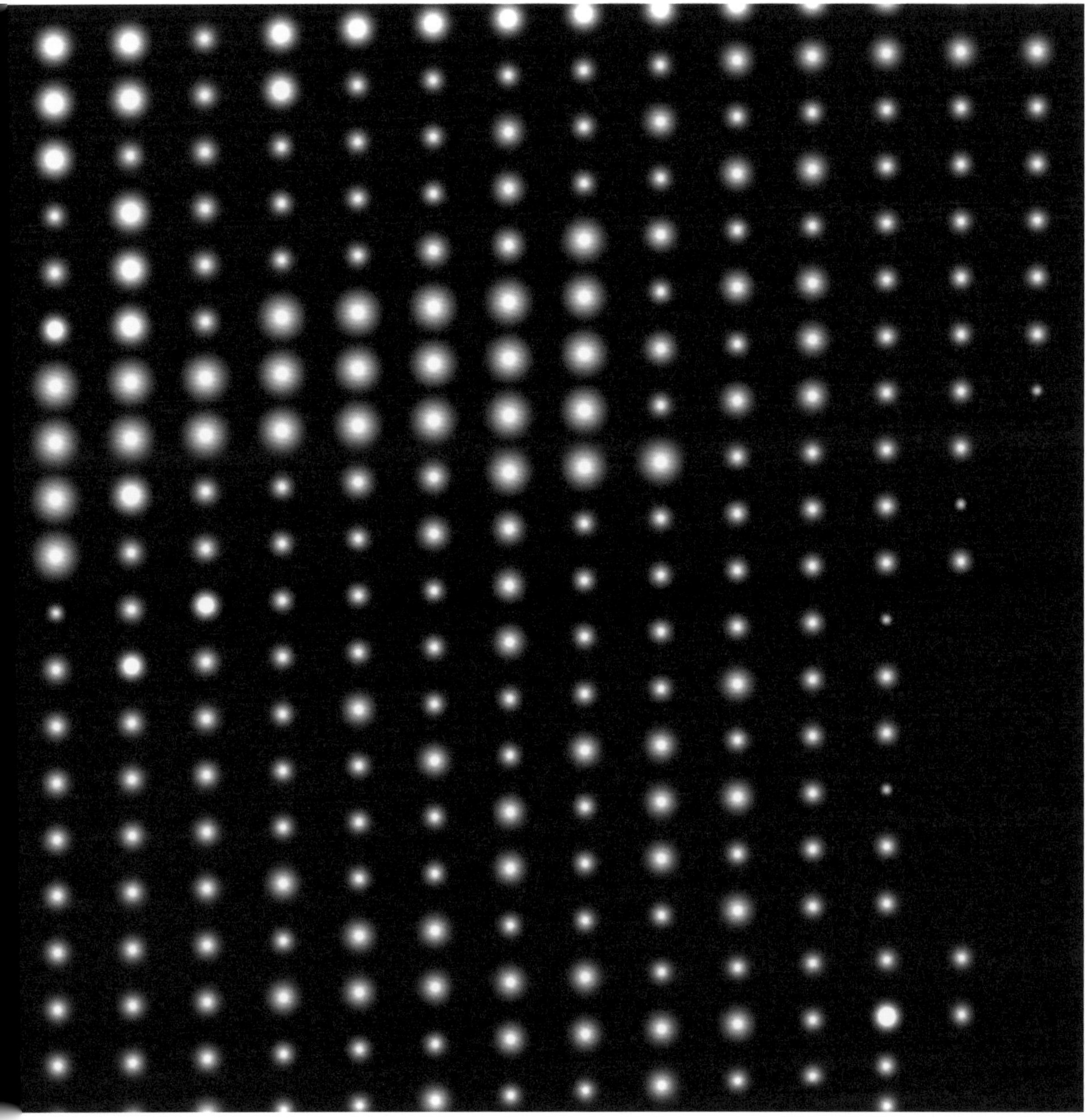

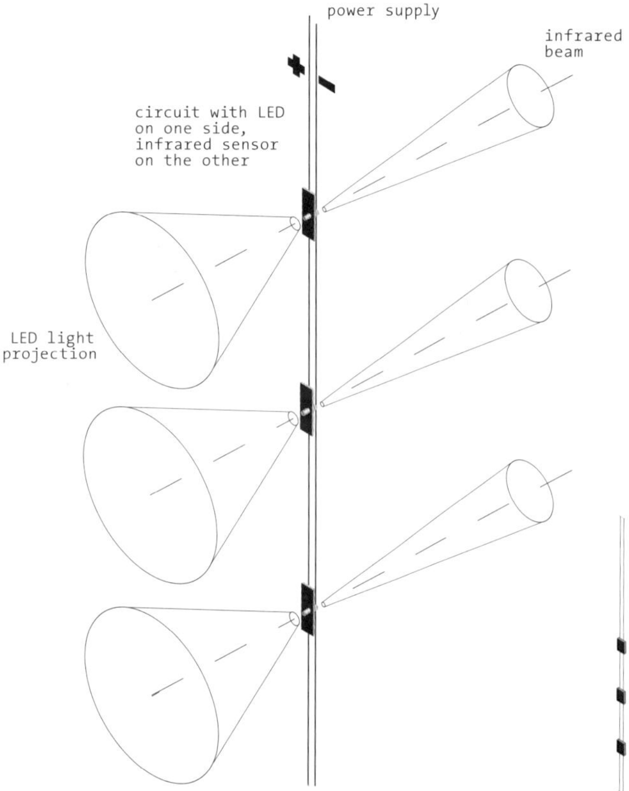

power supply
infrared beam
circuit with LED on one side, infrared sensor on the other
LED light projection

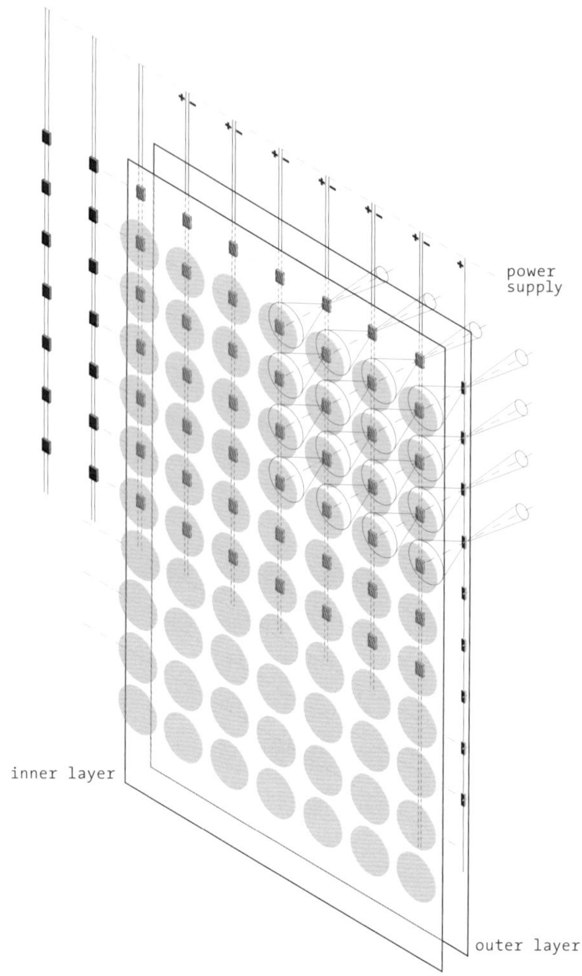

power supply
inner layer
outer layer

/02

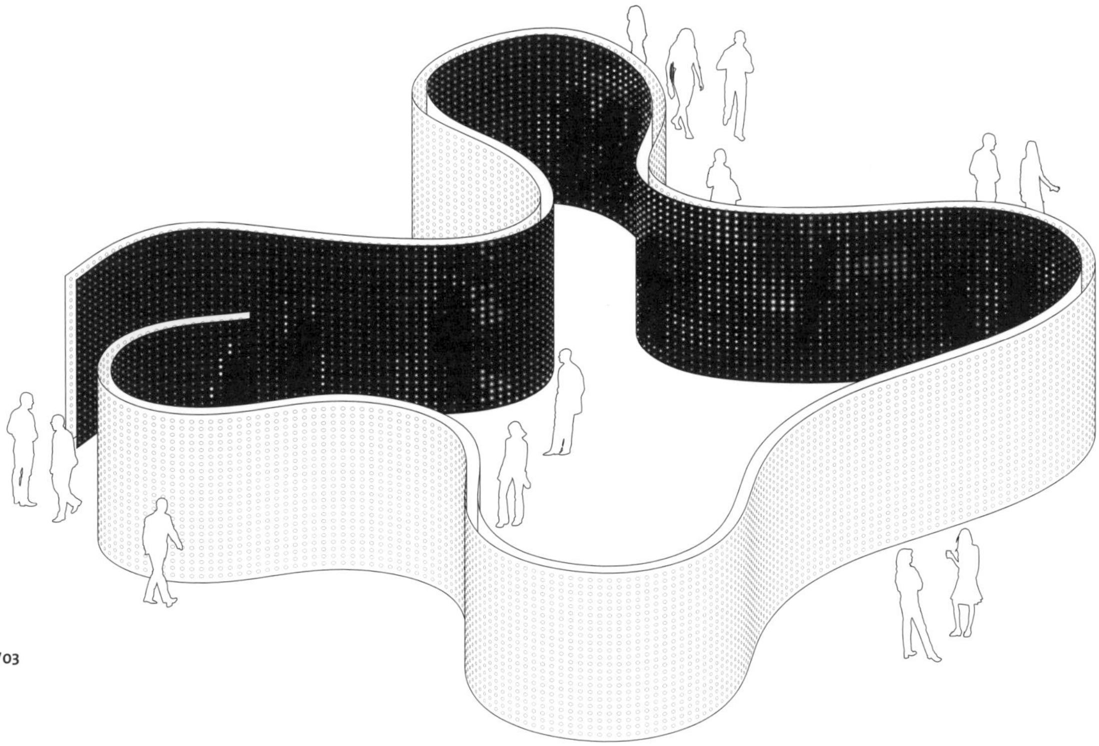

/03

/02
The curtain is composed of two separate layers. The outer layer is the supporting structure for the infrared sensors and the LEDs.
The inner layer forms the screen upon which the LEDs project light. When an object interrupts the infrared beams, the corresponding LED lights dim, causing a dark shape to emerge on the inner layer of the curtain.

/03
Through the use of electronic noise cancellation technology and sound absorbing materials, the enclosed space is almost silent.

dual state room

A space that responds to fluctuating light levels by continuously modulating its surface to enhance light and mood / **Studio NMinusOne**

The Dual State Room is a systematically constructed deep infrastructure for controlling the most superficial of tectonic conditions: the wall surface.

The Dual State Room can change its appearance. One moment it is a warm, sound-absorbent walnut veneer, and the next, a cool, reflective fluorescent yellow enamel. The space flip flops between these states, responding to light level fluctuations over the course of the day. When the space is too bright, the wall adopts the wooden side, reducing glare, and offering a feeling of warmth. Conversely, when dark, the wall flips to a highly reflective surface maximising light levels in dark zones. The Dual State system takes on the critical role of light modulation.

Its cellular construction – the result of the re-appropriation of discarded technology – is capable of responding locally by changing its material qualities depending on the weather, altering in turn the perception of the inhabitant.

/01
On a physical level, the wood veneer absorbs both light and sound. On a psychological level it offers a feeling of warmth and calm adopting a Martha Stewart-like interior design rhetoric.

/02
Bathed in yellow, the subject's mood is altered.

/01

/02

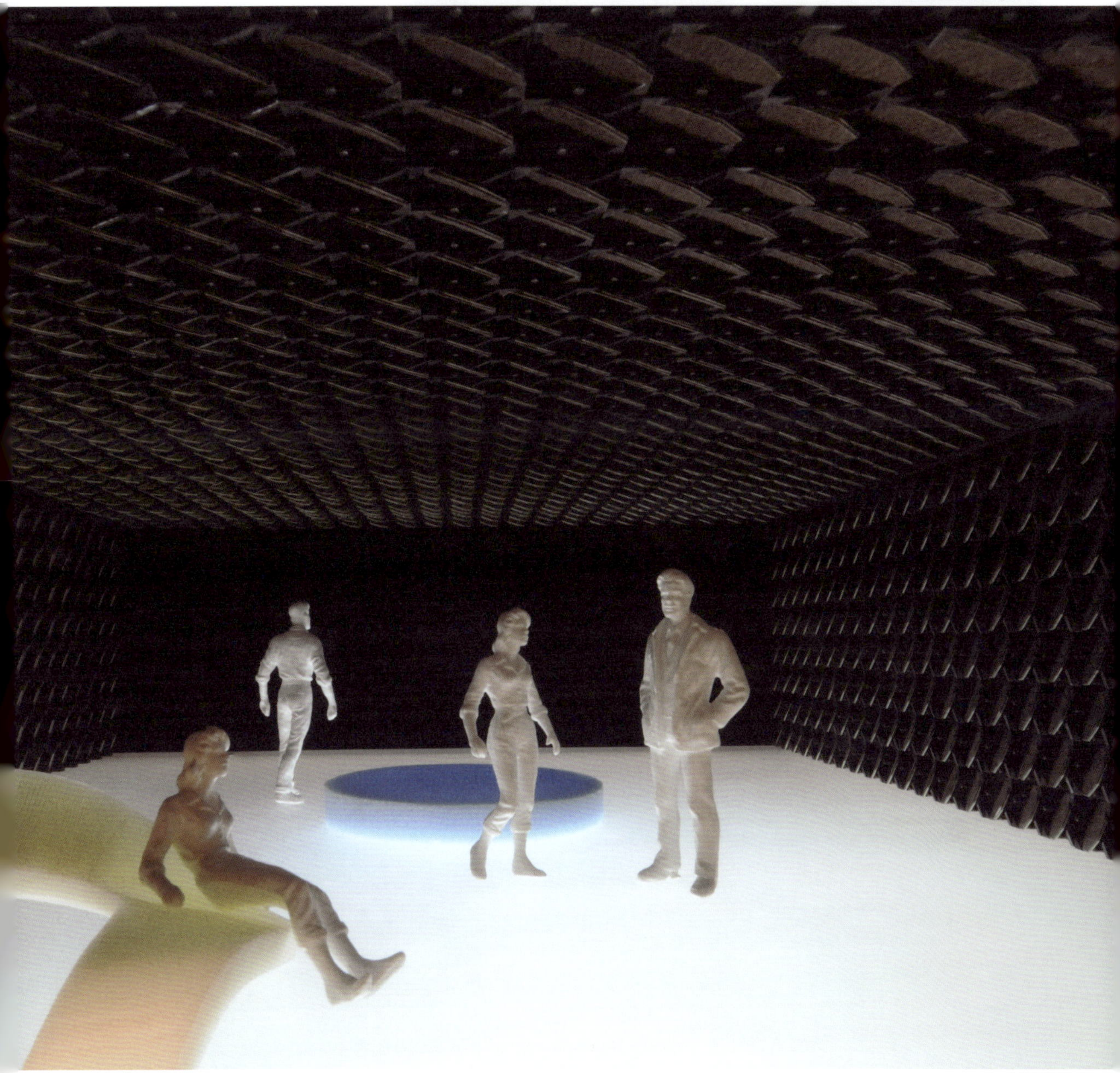

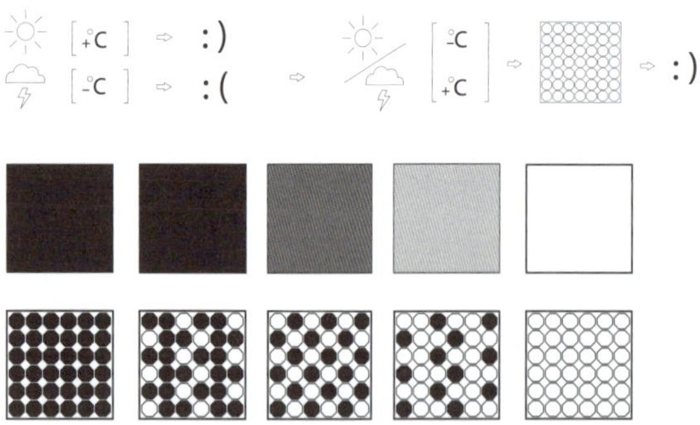

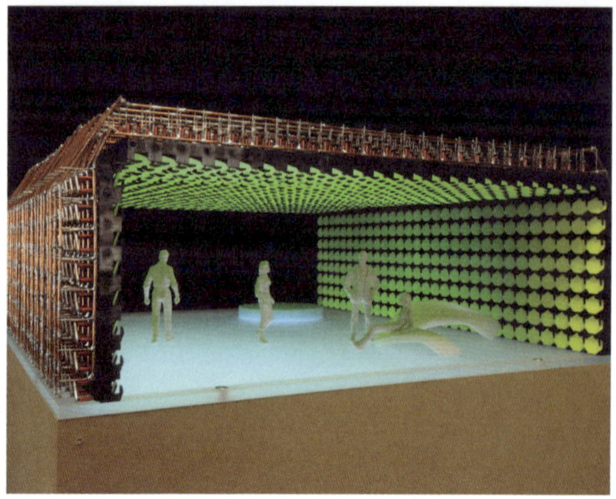

/03

/04

/03
The Dual State Room exploits the relationship between weather, interior light levels, and mood.

/04
Detail of a working prototype.

/05
Each cell operates autonomously, containing exterior and interior temperature and light sensors. The cells respond to fluctuating light levels, and change sides accordingly.

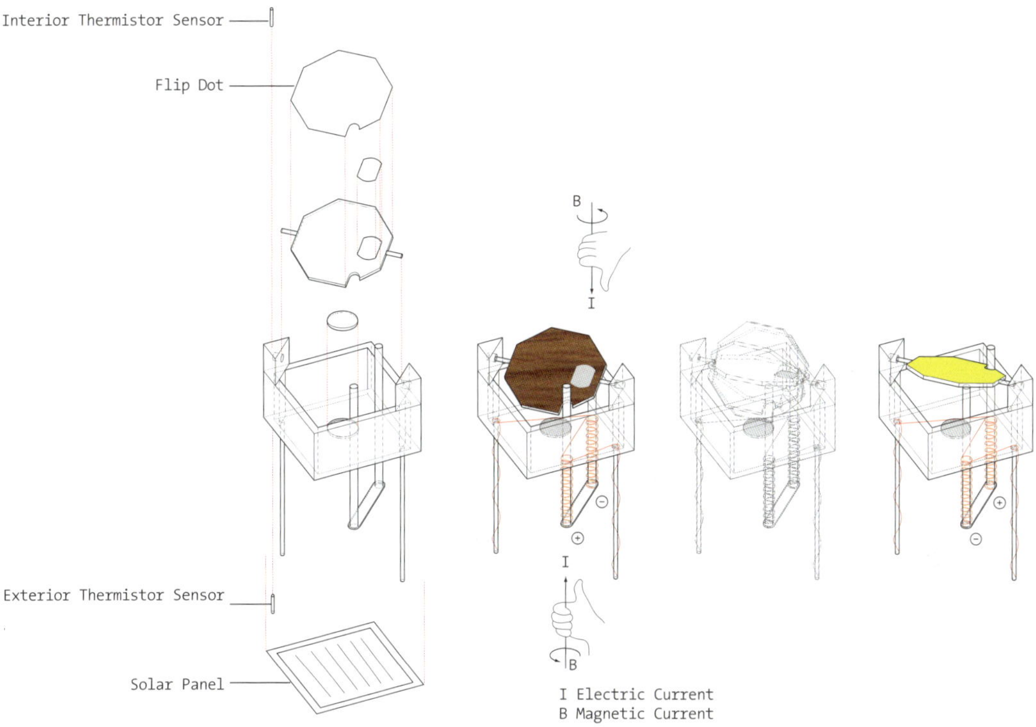

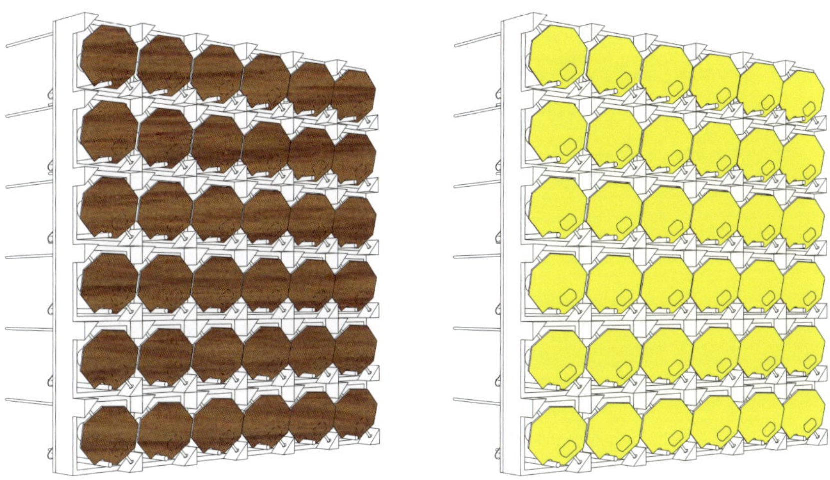

//Surface A//

Material: Wood
State: Max Absorption

//Surface B//

Material: Bright Fluorescent Paint
State: Max Reflectivity

xi’an scent garden

The Xi’an Scent Garden marries nature with artifice through a combination of fragrant local plant species alongside motion sensor-activated scent poles / **Khoury Levit Fong with Drew Adams, James Dixon and Fadi Massoud**

The Xi’an Scent Garden, commissioned for the 2011 Xi’an World Horticultural Exposition, delights all the senses while favouring smell.

Its terraced pine groves and beds of rosemary and thyme build and sustain a unique olfactory world. The local conifers create an olfactory, visual, and tactile connection to Xian. Motion sensor-activated scent poles diffusing plant extracts complement these fragrances with exotic accents evoking other locales to create an olfactory map of China. The constellation of glowing scent poles creates an iconic image of the garden as a public space, while its nuanced sensory output provides sheltered and individual experiences.

The scent poles double as light fixtures. They consist in stainless steel tubes with integrated scent cartridges and LED lights. The scent is diffused naturally by the breeze or, more promptly for approaching visitors, by a sensor-activated fan. The light intensity varies with the velocity of the wind. It is modulated by a wind sensor at the top of the pole. The lighting’s dynamic response translates the scent propagation pattern into a visual phenomenon.

The collection of artificial scents are bottled and packaged for sale to visitors in a glass pavilion.

/02

/01 Night time perspective view.

/02 View of main path.

/03 View of glass pavilion.

/04 Scent tube prototype.

/01

/03

/04

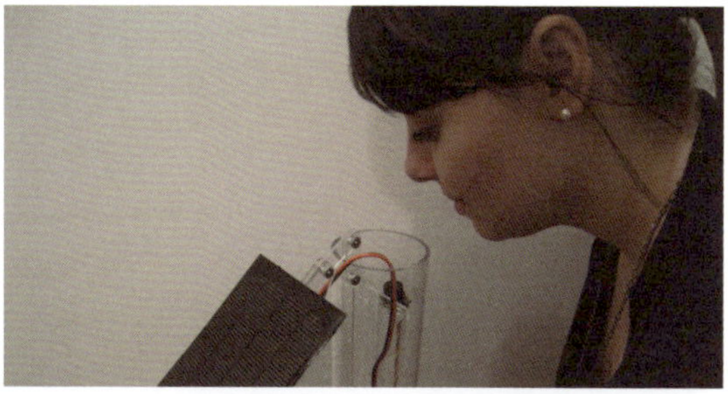

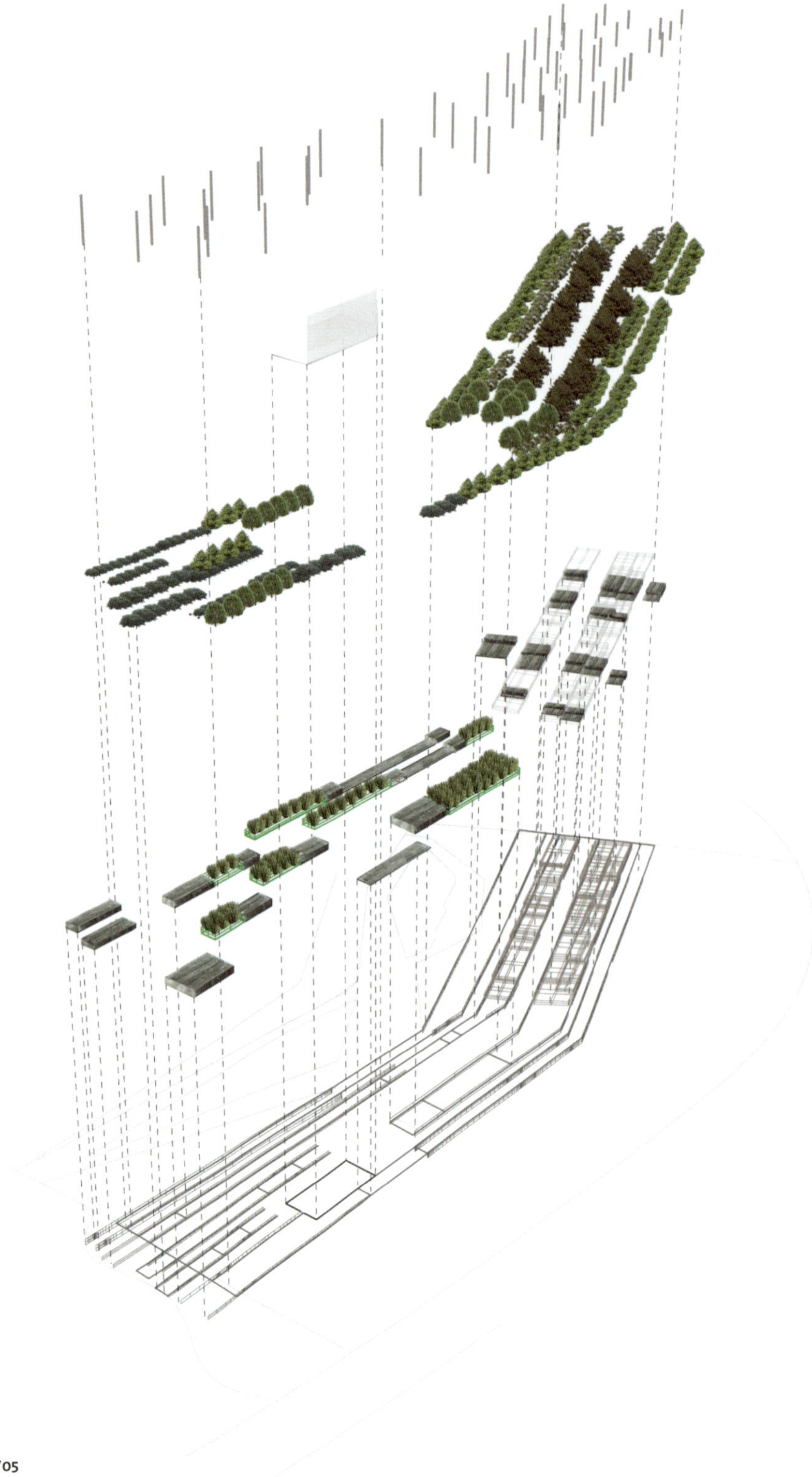
scent poles
pavilion - shop
pine and juniper grove
masonry seating and steps
and herb beds
steel dividers and terracing

/06

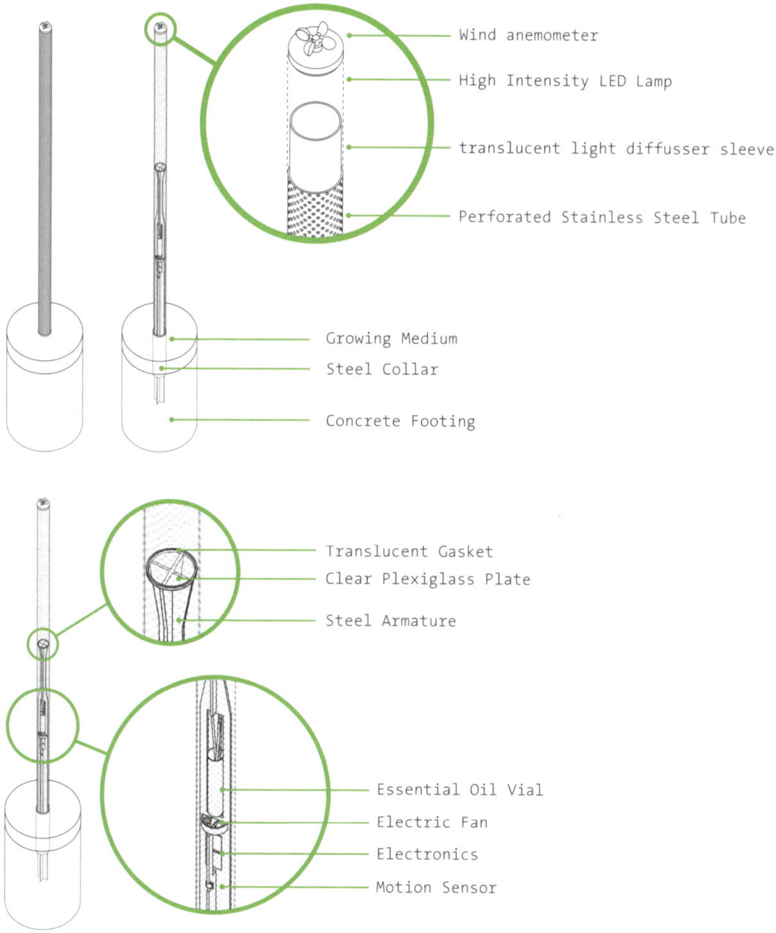

/07

/05
Exploded axonometric drawing showing the layering of different systems.

/06
Perspective view from hill.

/07
Digital scent pole assembly.

monuments + bits

The exhibit explores two strategies for the digital enhancement of physical objects / **Khoury Levit Fong**

The Monuments + Bits exhibition combines parametric modelling and fabrication techniques with emerging responsive media to present architecture and urban design projects in an interactive and digitally augmented environment.

The exhibition opened at the University of Toronto's Eric Arthur Gallery in September 2009 and travelled to the CMYK Gallery of the University of Michigan's A. Alfred Taubman College of Architecture and Urban Planning, running until the end of December 2009.

Monuments + Bits dramatises the event of collective monuments and their tense relationship with the immediacy of individual experience. Conventions are recast through new parametric capacities in the production of architectural form, while current developments in interactive media refashion perceptions of the fixity of the built environment.

The exhibition design takes advantage of the different characters of the Eric Arthur Gallery's two main spaces to explore contrasting strategies for the integration of digital technology and imagery into physical environments: while electronic and AV equipment is prominently displayed as part of a space-defining superstructure in one room (room A), it is seamlessly embedded in the other (room B).

/01

/01
View of AR station in foreground and responsive projection behind.

Monuments
+
Bits
WORK BY KHOURY LEVIT FONG
21 September – 31 October 2009

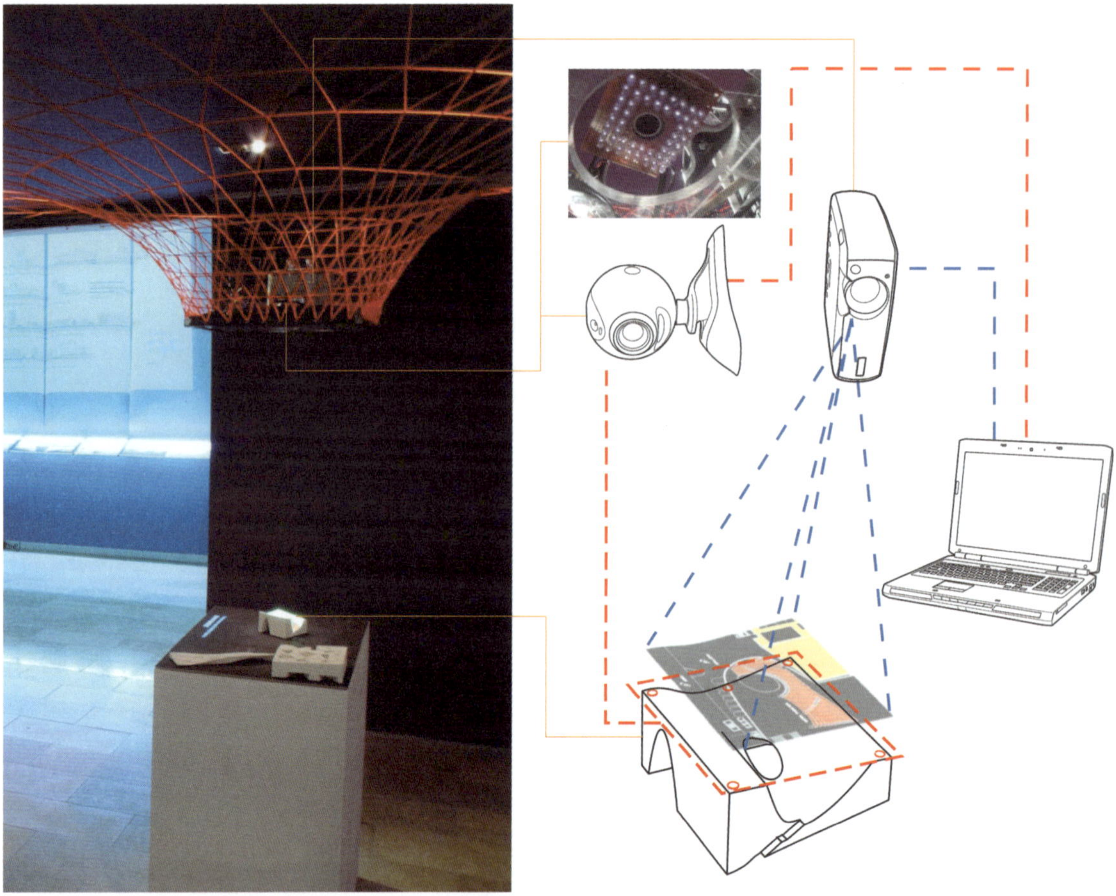

/02

/03

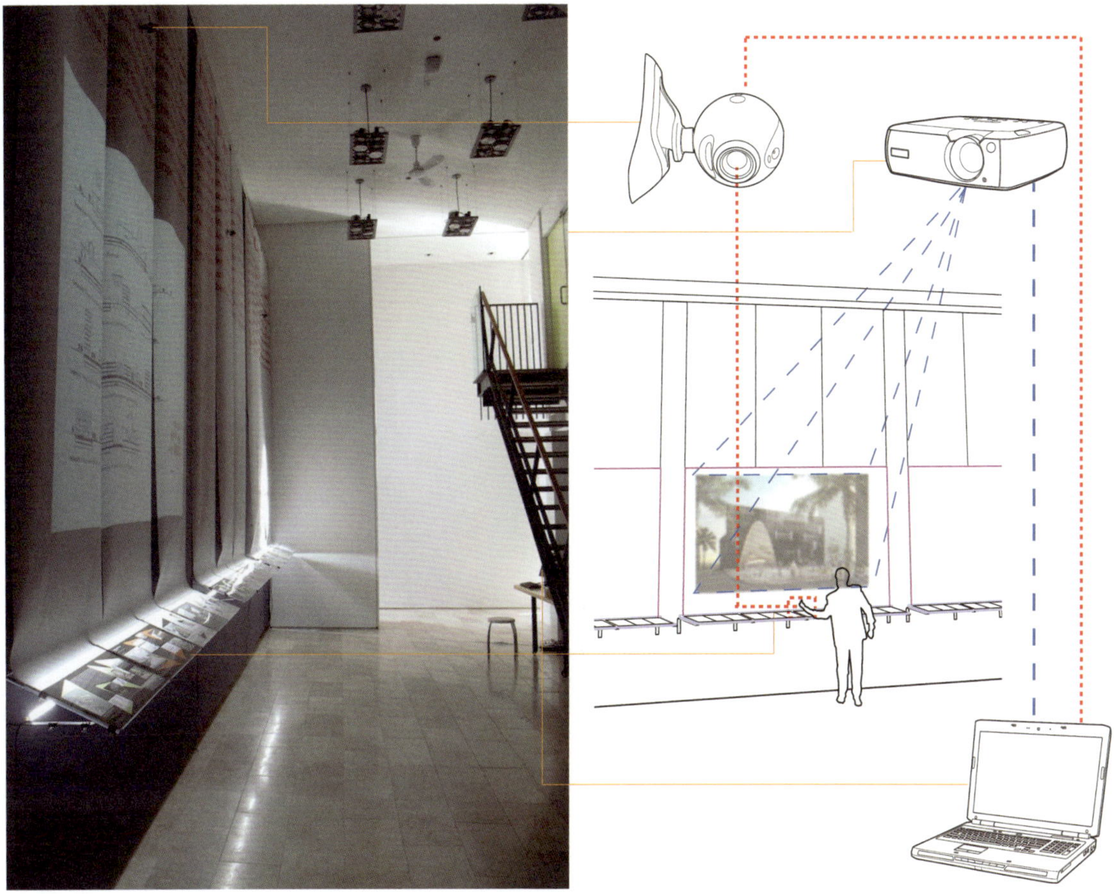

/04

/02-03

AR station. The system recognises the physical model and projects a digital animated image for additional information. The projection locks onto the selected objects and follows them within the projection area.

/04

The system recognises and tracks the visitors' pointing finger to select and project pertinent images complementing the project on display.

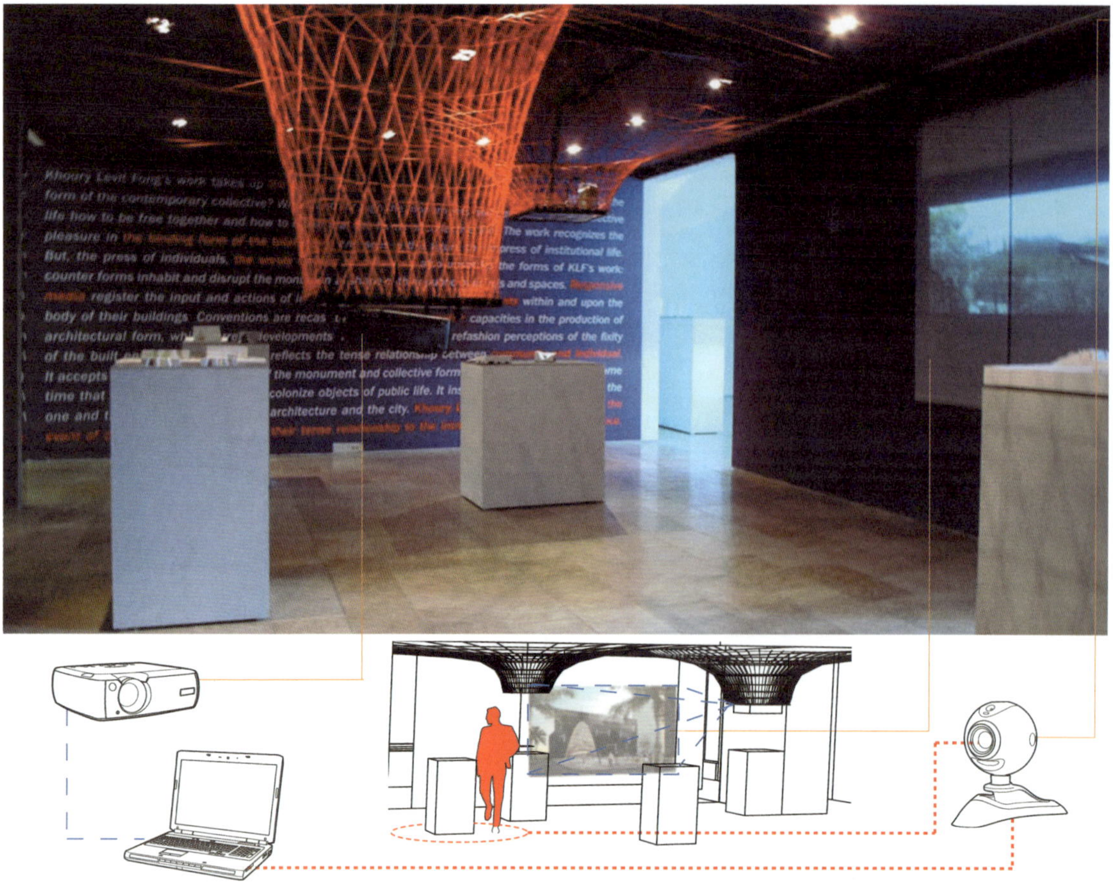

/05

/05
The system tracks the position of the visitor to display the images related to the model he or she is currently viewing.

NORDHAVNEN SUSTAINABILITY CITY
NORDHAVNEN, COPENHAGEN
MPETITION ENTRY, 2006
VIT FONG IN PARTNERSHIP WITH BROWN + STOREY ARCHITECTS
ALS IN CHARGE: R. EL-KHOURY AND R. LEVIT
ALS IN CHARGE: JAMES BROWN + KIM STOREY

blink

Concave eye-shaped surfaces reflect light upon pipelines, heating their contents and powering electric turbines housed beneath their crests / **Khoury Levit Fong**

A constructed dunescape with southern slopes of mirror-bright polished stone adapts the machinery of solar power generation, embedding it within an artificial landscape.

A rich pattern of mirrored eyes is luminous with the sun by day and shifting patterns of LEDs by night. The dunescape blinks.

Blink adapts the machinery of solar power generation. The machine does not sit in nature; rather, the dunes look natural but are a work of geometry and artifice. Biomorphic eye-like surfaces blinking in the dunescape uncannily animate the inanimate.

These dunes are stabilised: an open mat of concrete, mostly below the surface, holds their form and patterns their surface. Concrete ribs and planks form the substrate for the parabolic curvature of the mirrored surfaces. Under several of the larger dunes – over ten metres in height – powerful electrical turbines are housed. The mirrored surfaces reflect intense desert sunlight onto the pipes which carry superheated liquid to drive electricity-generating turbines.

At night, Blink turns into an informational ornament. The trajectory of air travel between Abu Dhabi and international destinations is mapped as trailing light patterns upon the mirrors of the dunes. This new electrical landscape signals the convergence between sustainable energy production and environmental art.

/01
Dunes at sunset: the dunes rise and subside, making for a richly varied array of forms, both to look at and to occupy.

/02
The solar dunescape is designed to be experienced at different speeds, at different scales and in different media: from Google Earth, a speeding car, the window of an airplane, or by foot.

/01

/02

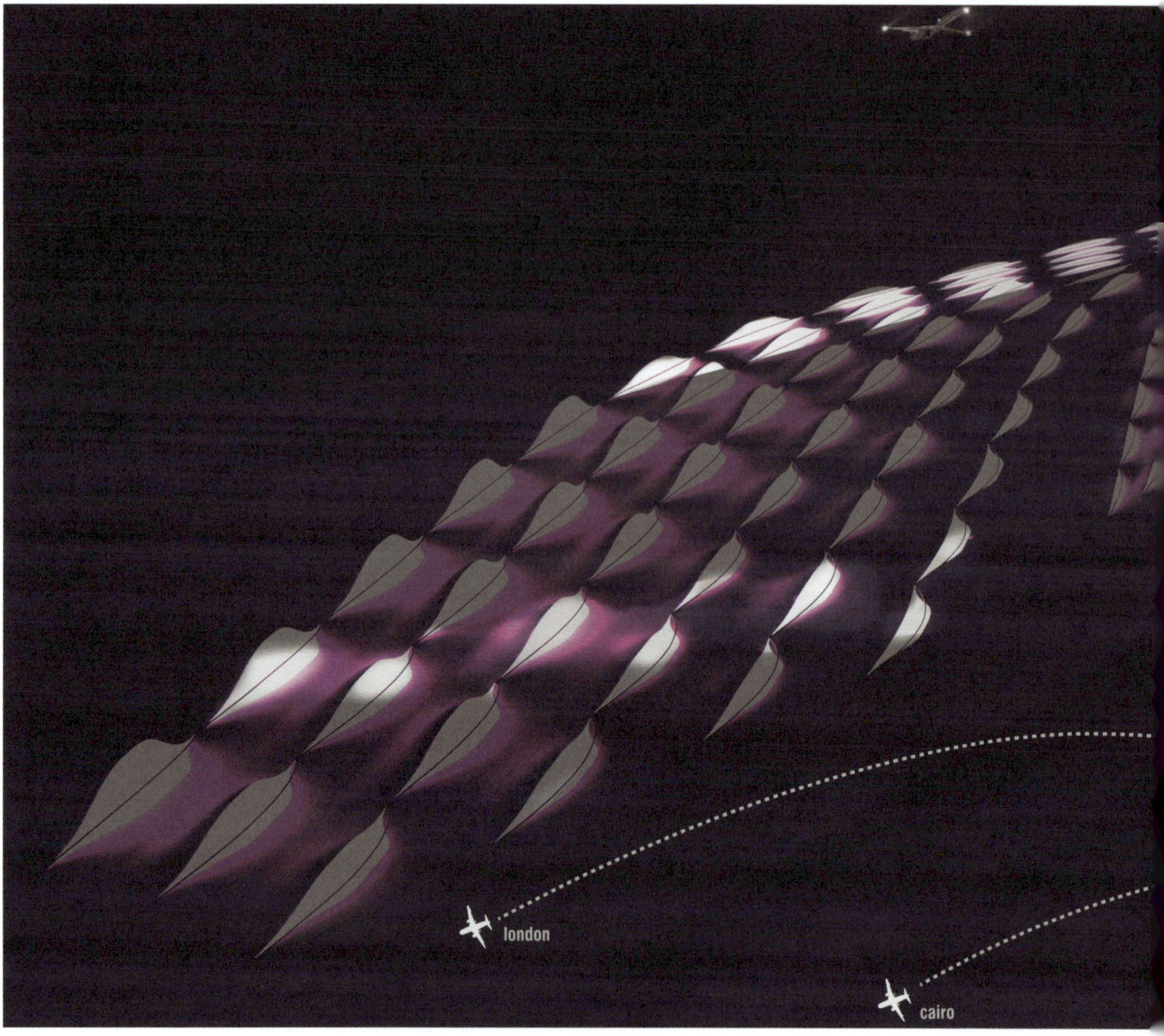

/03

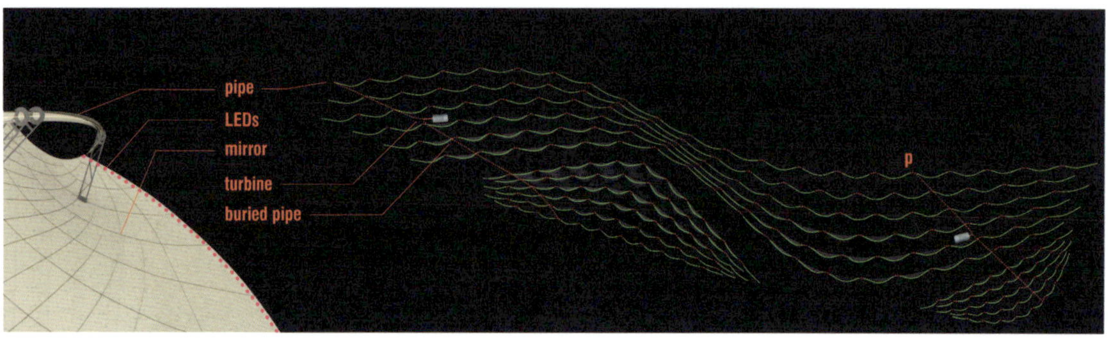

/05

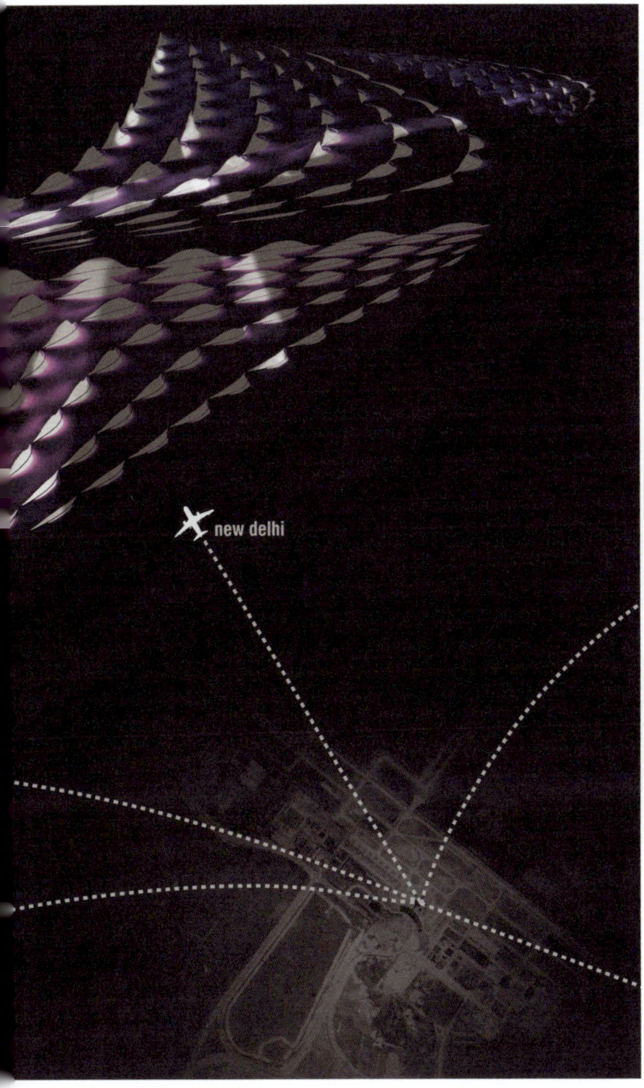

20:25’30 PM Arrival New Delhi

20:25’45 PM Arrival Cape Town

20:26’00 PM Departure London

20:26’15 PM Departure Cairo

/04

/03
LEDs embedded in the pipes in front of each mirror brighten and dim according to the arrivals and departures from Abu Dhabi airport.

/04
Electric generator turbine housed in a dune.

/05
Anatomy of a dunescape.

tunnel of light

A computer-controlled luminous shell envelops escalators at a design show and reacts to visitors' movement
/ **Khoury Levit Fong**

The two-layer tubular shell consists of nested rotational paraboloids whose lines are made from strands of fibre-optic filaments.

The complexity of a curvature that is made through the combination of straight lines makes for a visual and cognitive experience that permits immediate understanding while creating an ineffable effect that resists this easy grasp. This construct is animated with the application of proprietary responsive technology and techniques developed for constructing immersive and interactive environment by means of fibre-optic filaments. They serve to produce intricately shaped spaces and multi-sensory effects with quasi-immaterial means: ultra-thin cables and light.

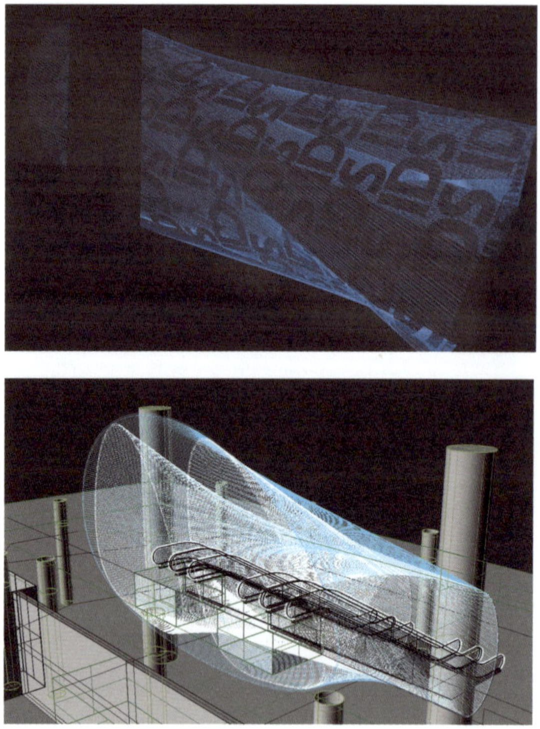

/01

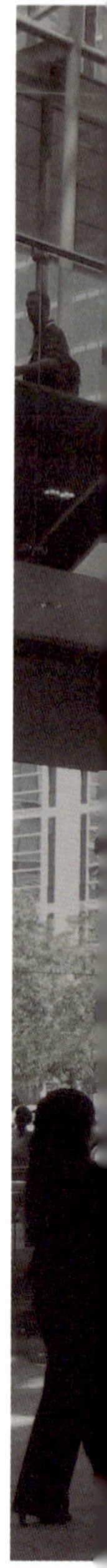

/02

/01
Parametric digital modelling of fibre-optic filaments.

/02
View of simulated effect.

/03

/03
Double-shell geometry study.

/04
Branding light pattern. The fibre-optic cables are treated for side-lighting in a designated pattern. This rendering shows the IDS acronym in a repeating pattern.

/05
The fibre-optic web is powered by micro-controlled lasers that interact with visitor's presence and movement. Responsiveness is calibrated by means of video-tracking devices. The light carried by the cables intensifies in a rotational sweeping motion. The greater the number of visitors, the faster the sweep.

/04

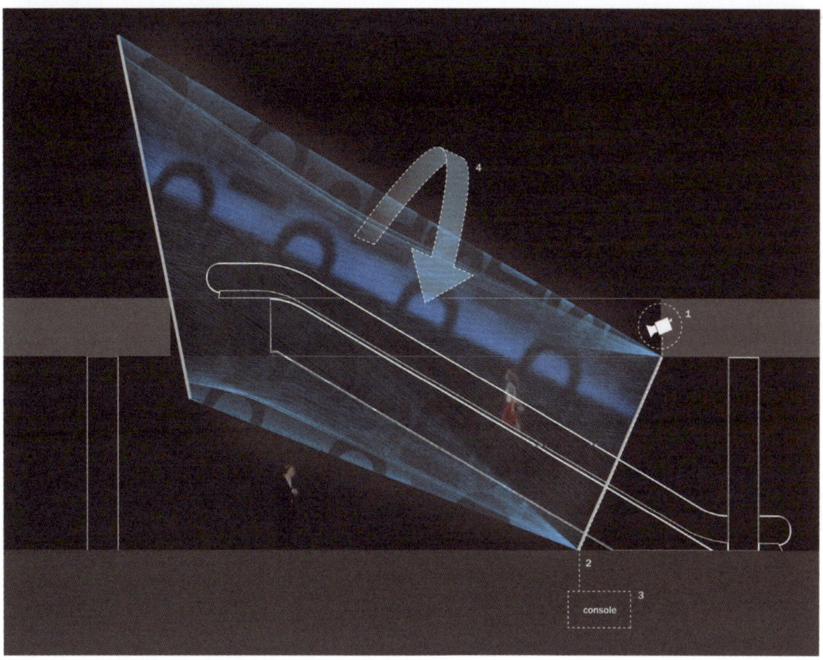

/05

rule

This exhibition of work by KLF at MIT demonstrates the phenomenal qualities of ruled surfaces used in recent building projects in an interactive installation of digitally controlled and laser powered fibre-optics / **Khoury Levit Fong**

The ruled surface exacts a discipline that relates architecture directly to the persistent modularity of building materials.

The ruled surface, which invites visual understanding and engagement, also dramatises the event of collective monuments and their tense relationship to individual experience in KLF's work. These forms, made of discrete parts, gather themselves up into figures that choreograph public experiences: they are both a framework and formal allegory for social event.

While geometry joins the pleasures of the intellect with those of the senses, explorations of responsive media and augmented reality are used to make immersive environments of light, image, and smell. New technologies mobilise the senses through media that are stimulated by the activities and presence of people. Scents are released and patterns of light are initiated in this exhibit in response to the presence of visitors. The immediacy of the olfactory, of the moving image and its response to the presence of individuals, draws together the scales of collective experience to the singularity of the individuals who are at its origin.

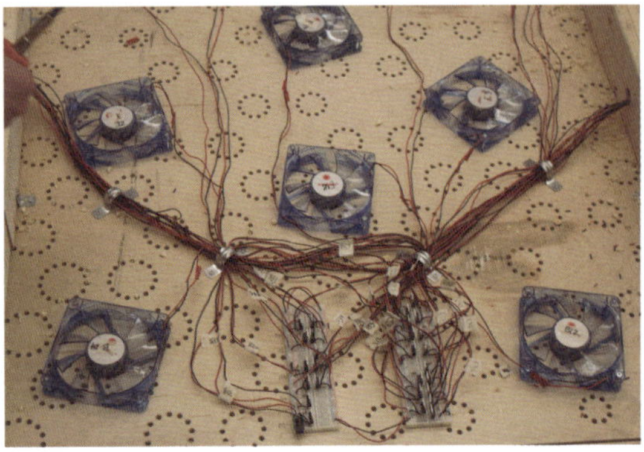

/01

/02

/01
Micro-controlled scent dispensers embedded in the perforated raided floor.

/02
View of the installation: a gossamer vault system made of rotational paraboloids.

/03

/03
Pattern of floor perforation for calibration of scent diffusion.

/04-05
Parametric digital model studies.

/06-07
Installation details.

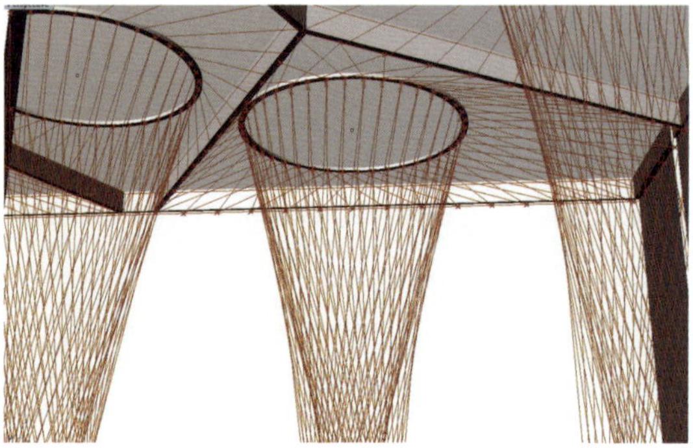

/04

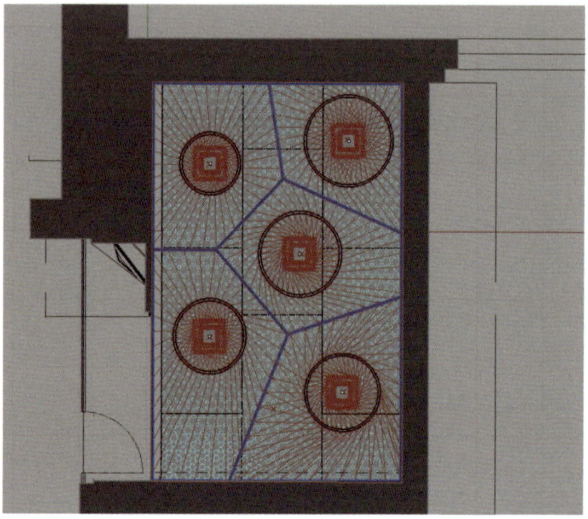

/05

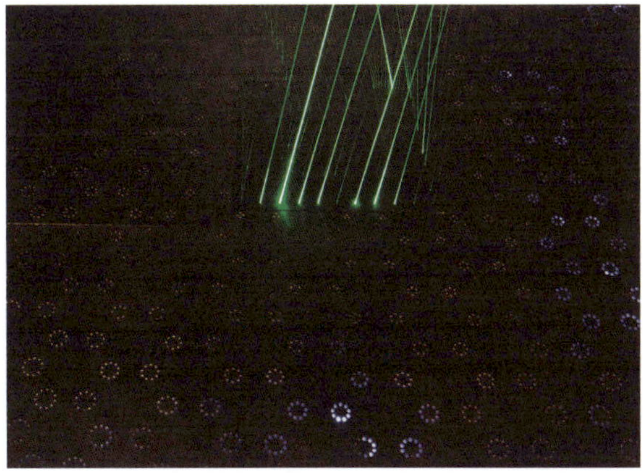

/06

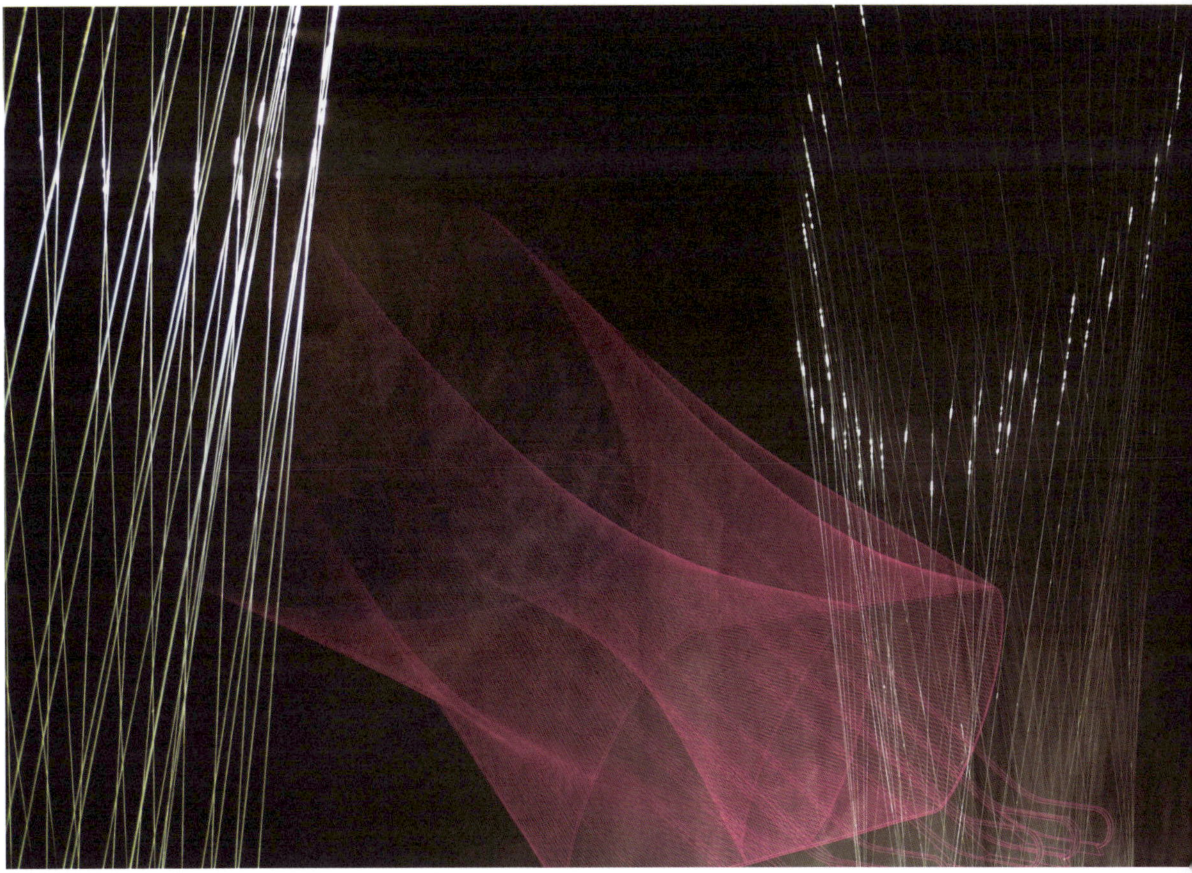

/07

nature 2.0

Flickering fibre-optic cables and scent-filled tubes imitate natural effects with the responsive light-burst of micro-controlled lasers
/ **Khoury Levit Fong**

Nature 2.0 consists of an artificial tree of fibre-optic filaments. Its backdrop is a luminous arboreal moiré of etched Plexiglas panels that gives the impression of a luminous electronic forest behind and beyond.

The intended effect is an immersive vision akin to that of being at the threshold of a forest: an interactive environment where digital technology reproduces the intricacy of sensory stimuli we associate with the teaming vitality of nature. The tree is shaped by strung fibre-optic filaments. They glow with the light they channel from a set of lasers into suspended glass tubes. The lasers flicker in patterns responsive to the proximity of visitors. These contain non-toxic and hypoallergenic chemical compounds designed by smell artist Sissel Tolaas that artificially construct an intricate olfactory image of nature. The artificial tree in Nature 2.0 is a geometrical construct – a rotational paraboloid with a diamond base and oval top (modified to fit the shape of our exhibition locations). This is a ruled surface artefact whose lines are made from 400 strands of fibre optic filaments. A canopy of filaments hang down from the oval top weighted at each of their ends by test tubes filled with coloured, scented oils. These test tubes hang in a stepped pattern that follows a similar geometry to that of the tree: a modified oval cut-out of a hyperbolic paraboloid.

The effect of this stranded form is to produce a glowing and scented environment in which the underlying energies of nature are echoed by the transmission of light through the fibre optic filaments. The floral scent of the tree is made artificially through the curated combinations of scents in the oils filling the test tubes.

/01
View of artificial forest.

/01

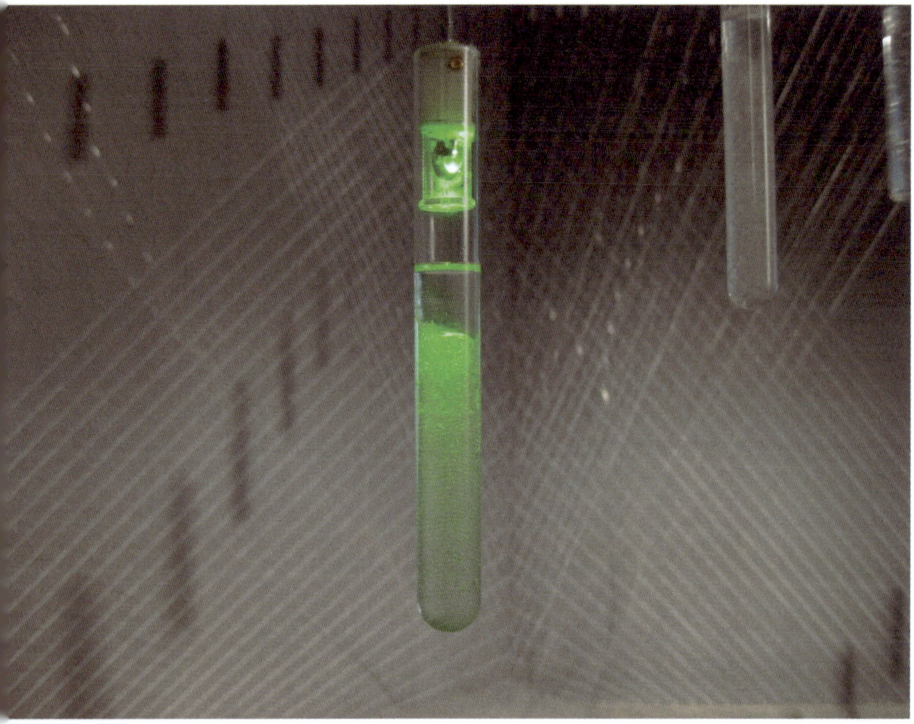

/02

/03

/02
Detail of scent diffusing tube with embedded LED and tilt sensor.

/03
Parametric digital model study for Toronto HarbourFront Centre version.

/04
Diagram of interactive features.

/05
Study for 2011 Chengdu Biennale version.

/04

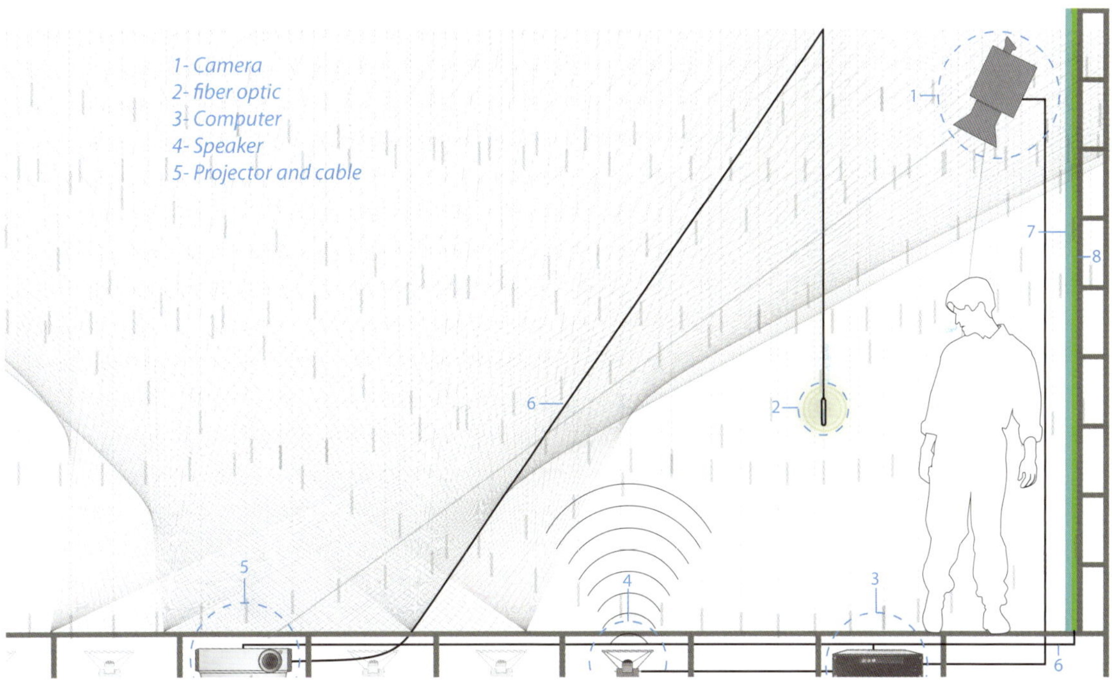
1- Camera
2- fiber optic
3- Computer
4- Speaker
5- Projector and cable
1
2
3
4
5
6
7
8

/05

/06

/06
Detail of rotational paraboloid geometry.

/07-09
Responsive lighting effects.

/10
Micro-controlled laser calibration.

/07

/10

/08

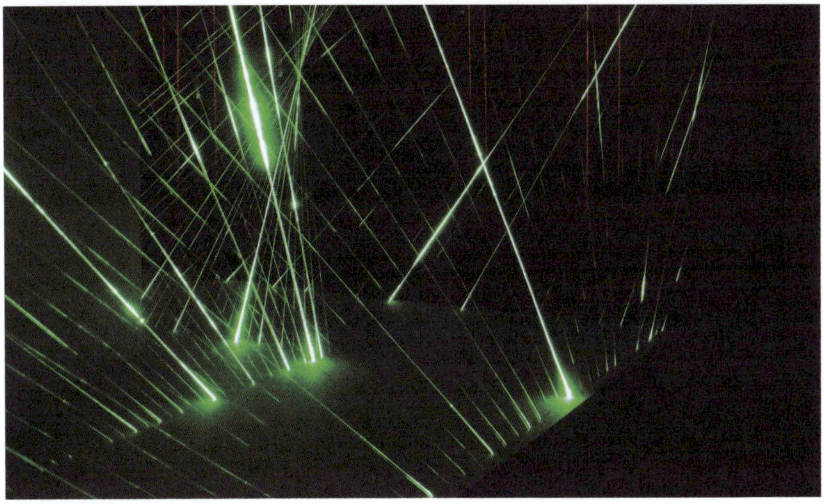

/09

Fog House

The electronically coordinated actions of misters and air curtains shape this wall-free house / **Ricky Ka Fung Luk***

Fog is the stuff of walls in this substance-free environment. It is produced by ultrasound misters and shaped by air curtains.

The Fog House consists mainly of two horizontal planes: a floor and a ceiling. In between are structural piers that accommodate mechanical equipment, fixtures and appliances. They anchor the loosely defined spaces with functional amenities. For instance, the pier that contains kitchen appliances anchors the kitchen area.

The different rooms of the house have no substantial enclosure. They are constituted by the computer-controlled and mechanically sculpted fog. The walls of any given room thus emerge or disappear according to needs and desires. Rooms are in a constant flux of formation and dissolution. They come into sharp focus on demand, their edges neatly defined by the air curtain, and dissolve into formless fog when not needed or seen.

The endless possibilities of responsive space-shaping scenarios include privacy that materialises only in anticipation of an approaching passer-by, rooms that expand or shrink according to need and compartments that materialise and open in response to a gesture.

/01
View of the Fog House in its pastoral setting.

/02
Air curtains carve out a doorway for an approaching user from the mister-generated fog.

*Guided by Rodolphe el-Khoury

/01

/02

Exterior

/03-05
Various scenarios for defining and controlling the fog-formed walls.

/06
Section and plan of Fog House.

/07
Diagram showing how rooms emerge in response to user input and movement.

/08
Machine for testing and calibrating the shaping of fog.

/03

/04

/05

/06

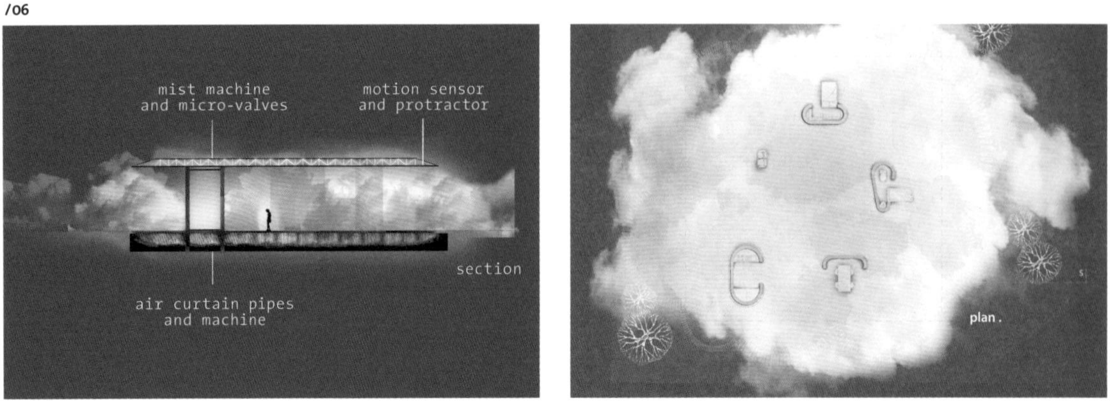

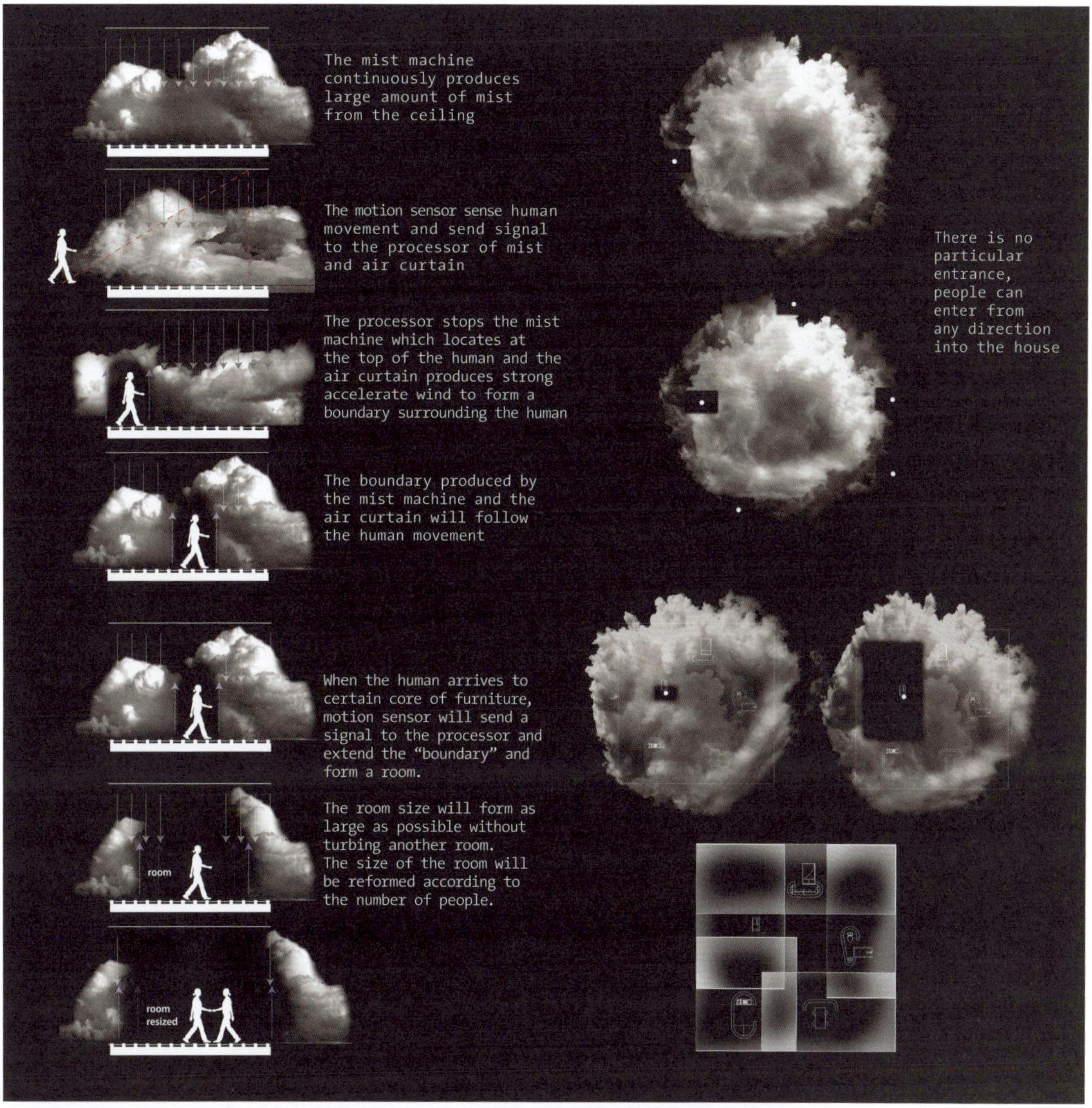
The mist machine continuously produces large amount of mist from the ceiling
The motion sensor sense human movement and send signal to the processor of mist and air curtain
The processor stops the mist machine which locates at the top of the human and the air curtain produces strong accelerate wind to form a boundary surrounding the human
The boundary produced by the mist machine and the air curtain will follow the human movement
When the human arrives to certain core of furniture, motion sensor will send a signal to the processor and extend the "boundary" and form a room.
The room size will form as large as possible without turbing another room. The size of the room will be reformed according to the number of people.
room
room resized
There is no particular entrance, people can enter from any direction into the house

/07

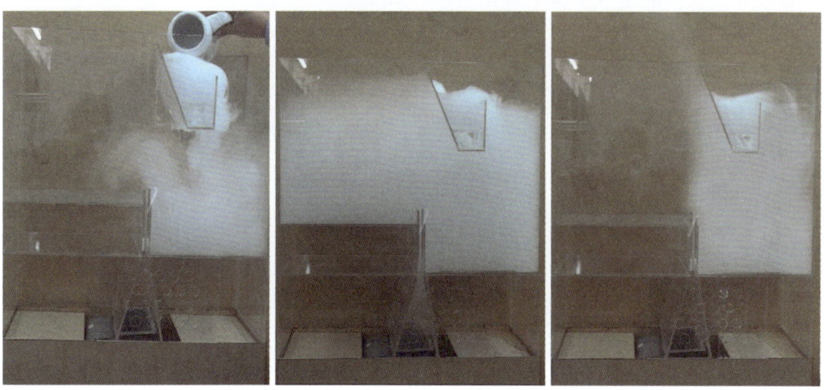

/08

kinetic components

The real-time sensing and actuating capacities of embedded systems enables a decentralised and networked organisation where individual building components adjust and adapt locally to environmental pressures, generating emergent global configurations. The distributed intelligence yields a more efficient and resilient system that is better equipped for managing environmental dynamics. The following projects rethink the building unit as an architectural quantum that is augmented digitally and capable of kinetic responses to dynamic internal and external stresses.

variable insulation

Insulation is released from the confines of the wall assembly, freeing it to adapt to varying thermal conditions / **Cleo Buster***

By relocating insulation from within the wall assembly to a flexible interior curtain, Variable Insulation is able to address shifting thermal needs as well as changing programmatic uses.

Two curtain prototypes were fabricated using two different materials. The first curtain is made of many individual aluminised Mylar pillows attached together. Each pillow is connected by polyethylene tubing to a centralised air pump and fitted with two-way valves. This enables the curtain to be inflated for increased insulation values, and deflated for decreased insulation values. The curtain responds to both exterior and interior temperatures, giving the inhabitant control over their environment.

The second prototype consists of individual cells constructed from one continuous piece of polyethylene tubing. Thus, the infrastructure becomes the structure, eliminating the need for two separate systems. Each cell has a clear latex balloon attached to a two-way valve, which is connected to the central matrix of the system. The balloon curtain has multiple functions: it simultaneously acts to insulate, mediate privacy and control light.

/01

/01
The curtain can achieve multiple R-values by altering the air space within each Mylar pillow. Each pillow is independently responsive to its surrounding ambient parameters.

*Guided by Christos Marcopoulos and Carol Moukheiber

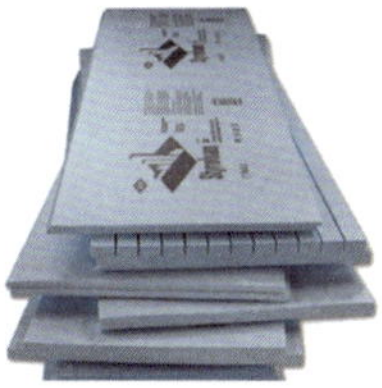

+

=

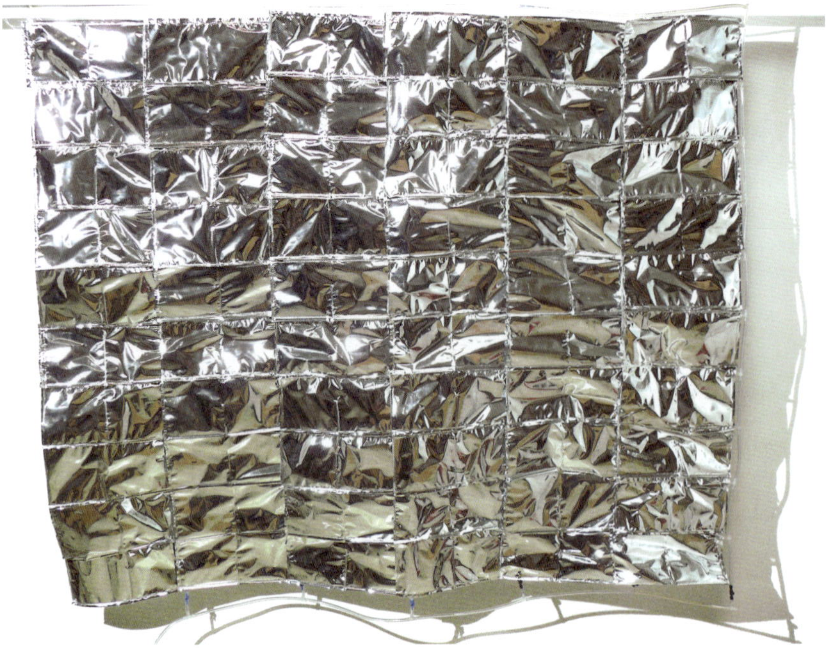

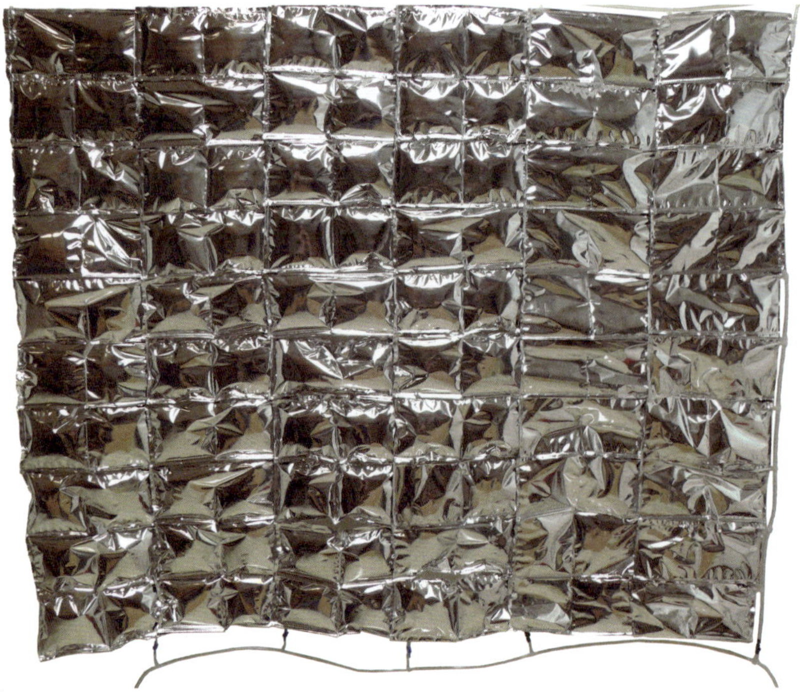

/02

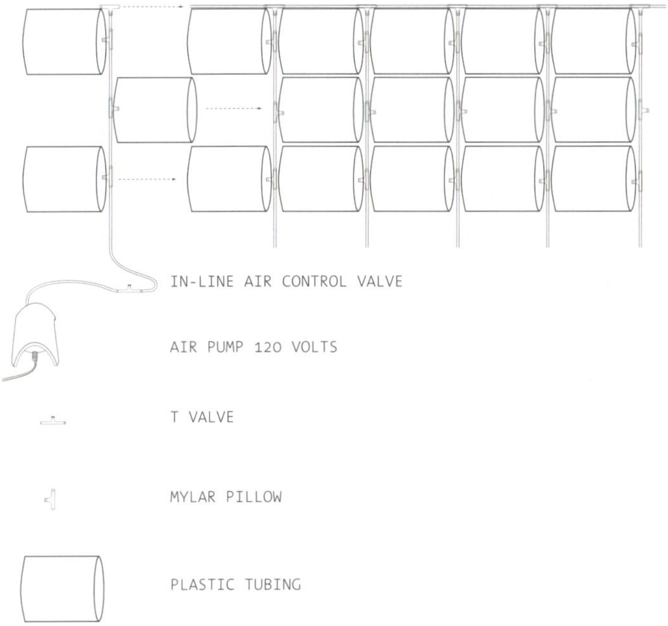

/03

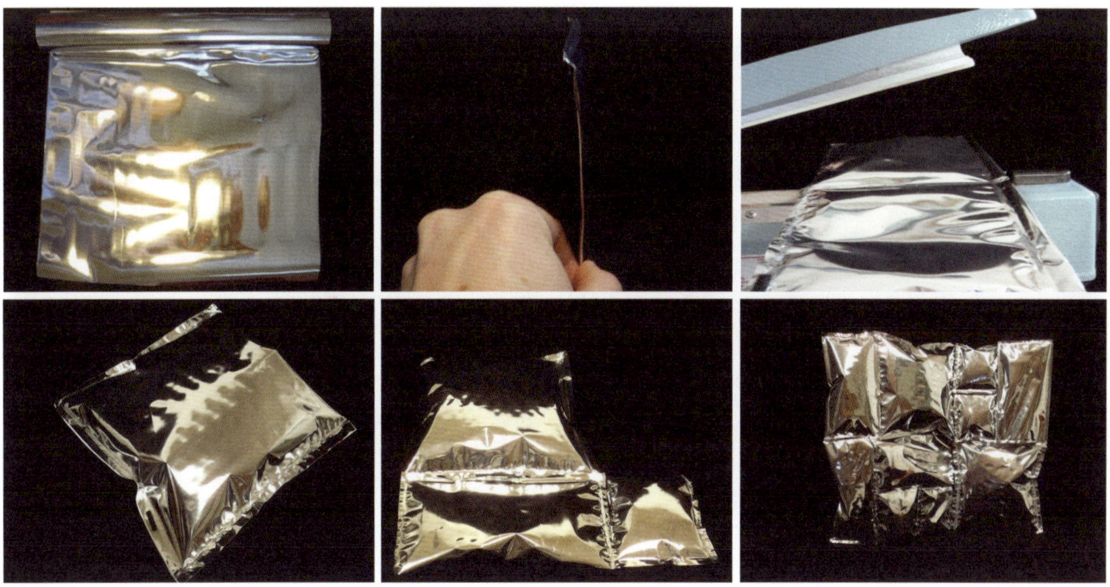

/04

/02
The curtain deflated and inflated.

/03
Each Mylar pillow plugs into the air supply infrastructure.

/04
The prototype was constructed using foil and a sealing machine from a commercial food packaging plant.

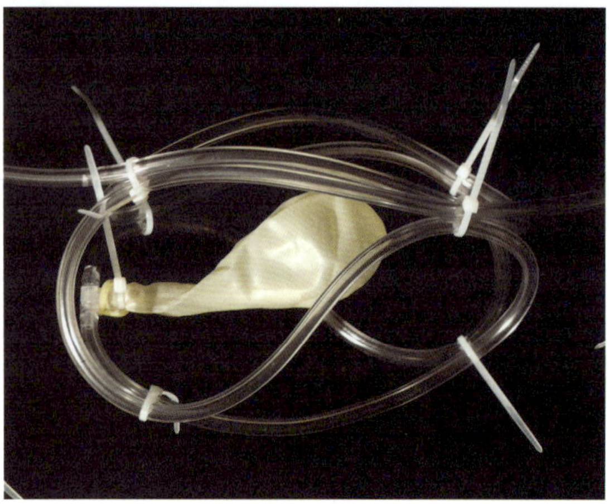

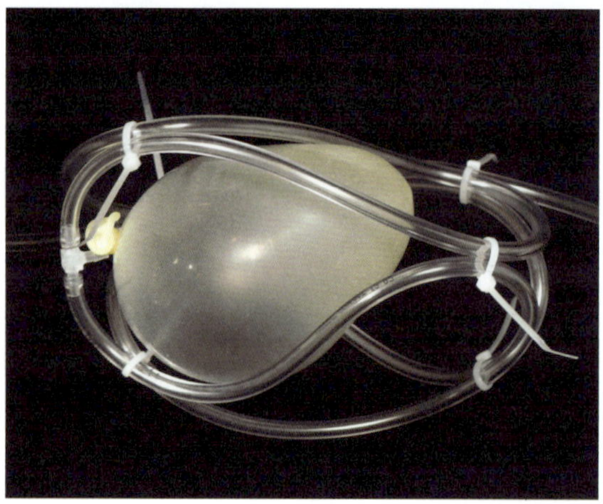

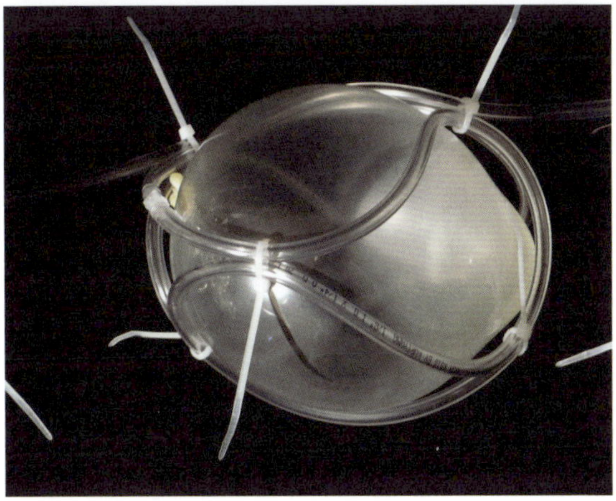

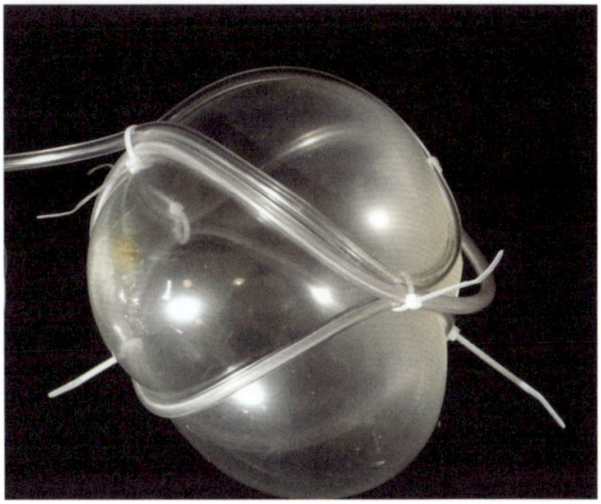

Using balloons and air tubing as both structure and infrastructure, this prototype explores the potential to change a surface's qualities: opaque and insulating vs. open.

fresh-air window

A sealed glazed curtain wall module is capable of supplying fresh air / **Studio NMinusOne with Valentina Mele**

The Fresh-Air Window is a sealed curtain wall unit with a Heat Recovery Ventilation element embedded in the double glazing, allowing incoming fresh air to be heated or cooled by the outgoing interior air.

Traditionally, heat recovery ventilators are centrally located. By contrast, the Fresh-Air Window decentralises the Heat Recovery Ventilation throughout the building. Through sensing devices, every window has the capacity to gauge interior air quality. When local air quality declines, the respective Fresh-Air Windows respond by introducing fresh air. Infrastructure is now embedded within the building block.

When your house contains such a complex of piping, flues, ducts, wires, lights, inlets, outlets, ovens, sinks, refuse disposers, hi-fi reverberators, antennae, conduits, freezers, heaters – when it contains so many services that the hardware could stand up by itself without any assistance from the house, why have a house to hold it up? – Reyner Banham, 'A Home Is Not A House'.

/01
The Fresh-Air Window employs two back-to-back Central Processing Unit cooling sinks. This configuration creates an instant Heat Recovery Ventilator.

/02
Traditional system: centralised heat-recovery ventilation; Proposed system: distributed heat recovery ventilation over the entire building envelope.

/01

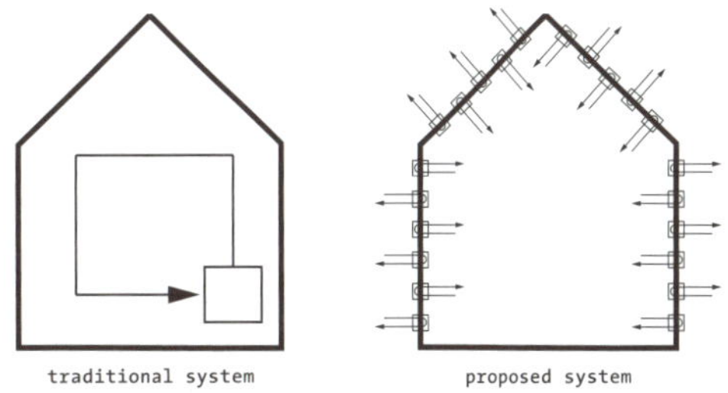

/02

/03

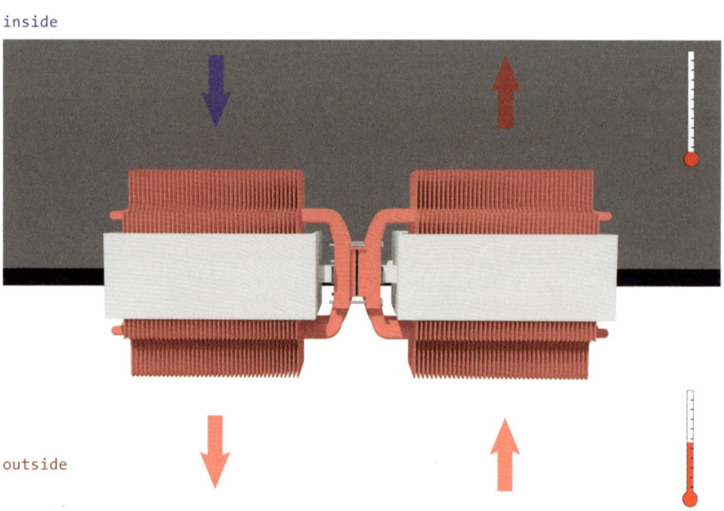

/04

/03
Two back-to-back CPU cooling units form a Heat Recovery Ventilator.

/04
Diagram of the Heat Recovery Ventilator: fresh cold air comes in from the exterior through the fins and is heated by outgoing warm, stale air from the inside going out.

/05
The window as air quality monitor.

/06
CPU HRV components: fans, copper tubing, and fins.

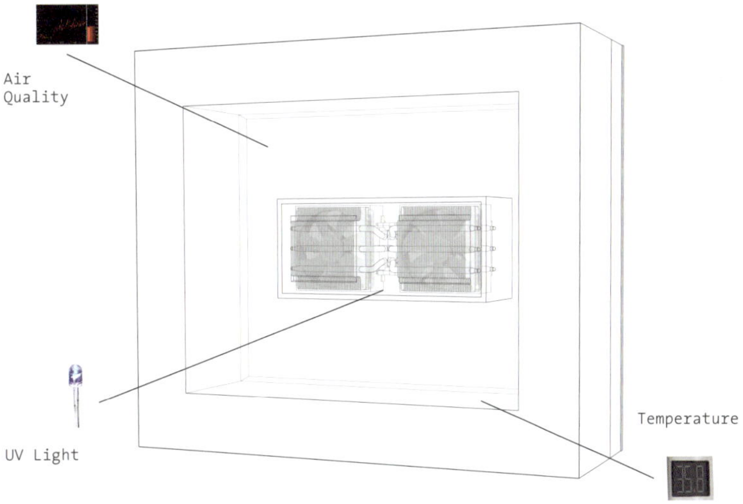

/05

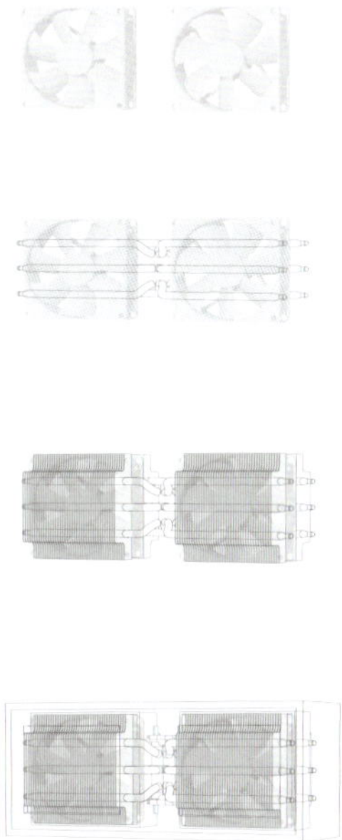

/06

heating and cooling glass

By distributing the previously centralised HVAC system into the glazing system, glass becomes a haptic transmitter / **Valentina Mele***

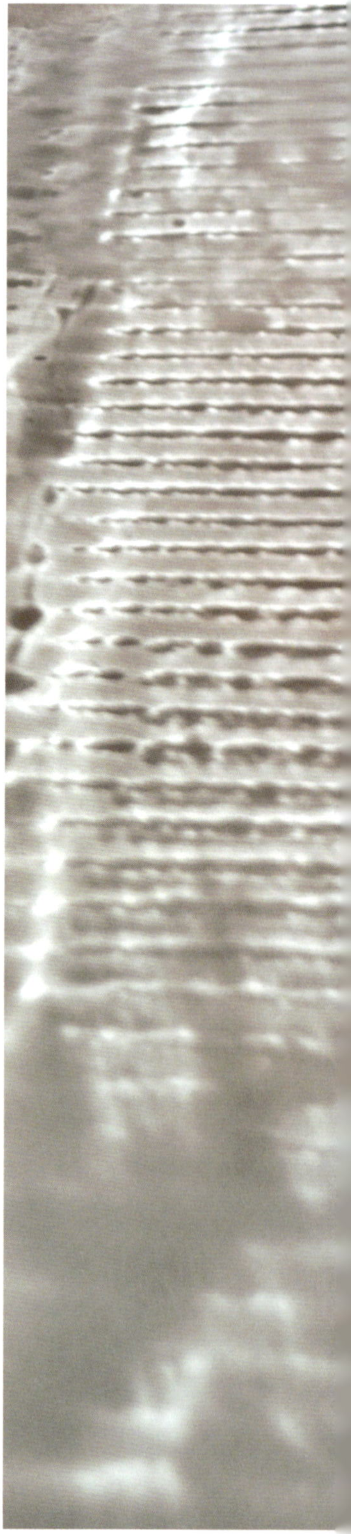

Heating and Cooling Glass explores the potential for the migration of the central HVAC system into a decentralised modular surface.

The double glazed Heating and Cooling Glass unit is used opportunistically as the site of maximum difference in temperature. Mediating incoming and outgoing air at the envelope takes advantage of the building's skin as the interface between two different climatic zones. That difference is harnessed – through a Peltier device inserted in the glazing unit –to generate heating and cooling locally, as needed by the user, or as necessitated by weather conditions.

By distributing the previously centralised HVAC system into the glazing system, glass becomes not only a visual transmitter of the outside but also a haptic transmitter. In a tactile way, it fulfils the quest of modernism to achieve continuity between inside and outside.

/02

/01
Shadow cast by the Heating and Cooling Glass.

/02
The miniaturised heating and air-conditioning unit consists of a Peltier device inserted between two back-to-back mounted CPU cooling sinks. The device cools and heats by reversing the current flow to the Peltier device.

*Guided by Christos Marcopoulos and Carol Moukheiber

/01

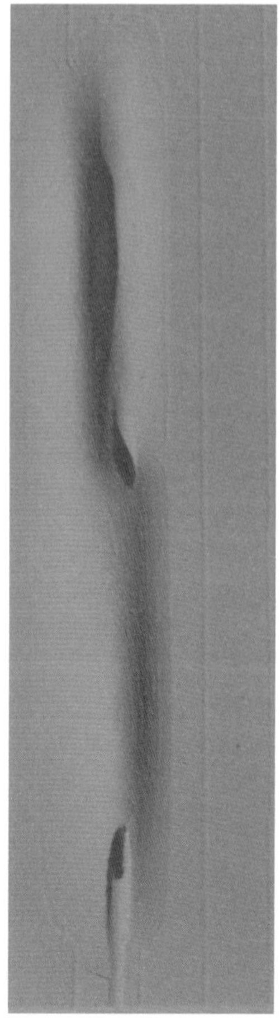

/03

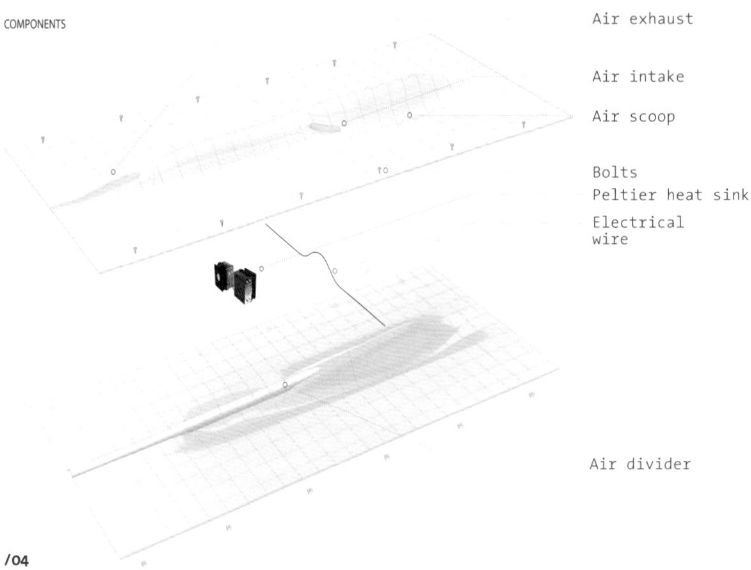
COMPONENTS
Air exhaust
Air intake
Air scoop
Bolts
Peltier heat sink
Electrical
wire
Air divider

/04

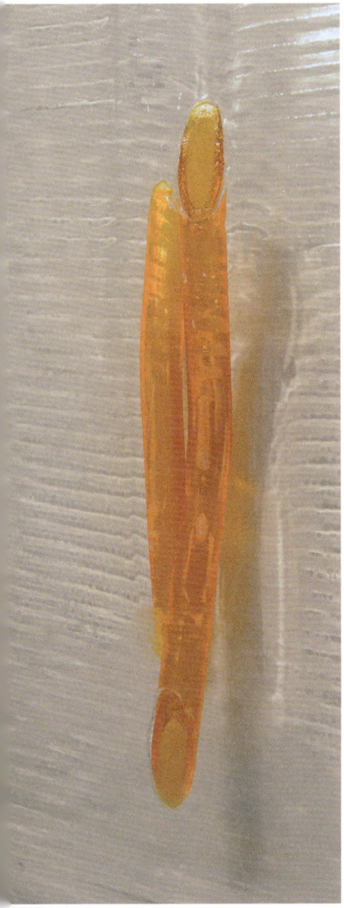

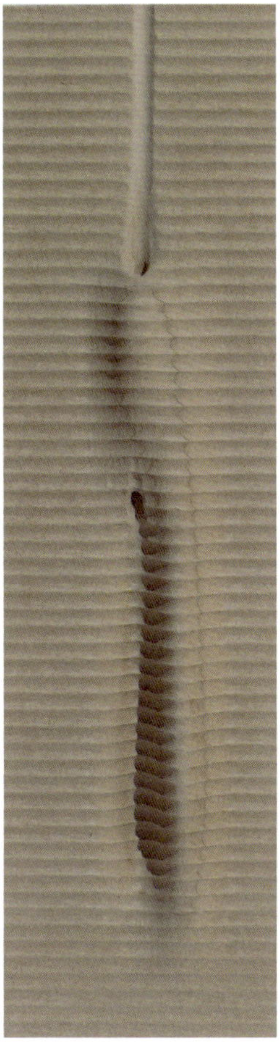

/03
The glazing itself is slumped to accommodate its new performance.

/04
The glazing sandwich. Two sheets of glass direct air in and out through the miniaturised heat pump.

/05
Diagram of the tubing infrastructure showing the airflow in plan section and elevation.

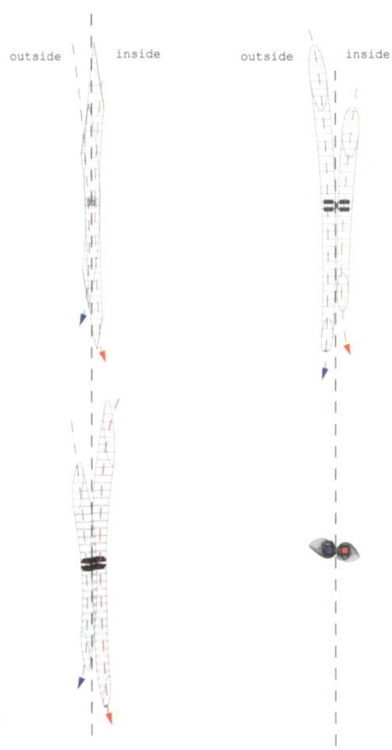

/05

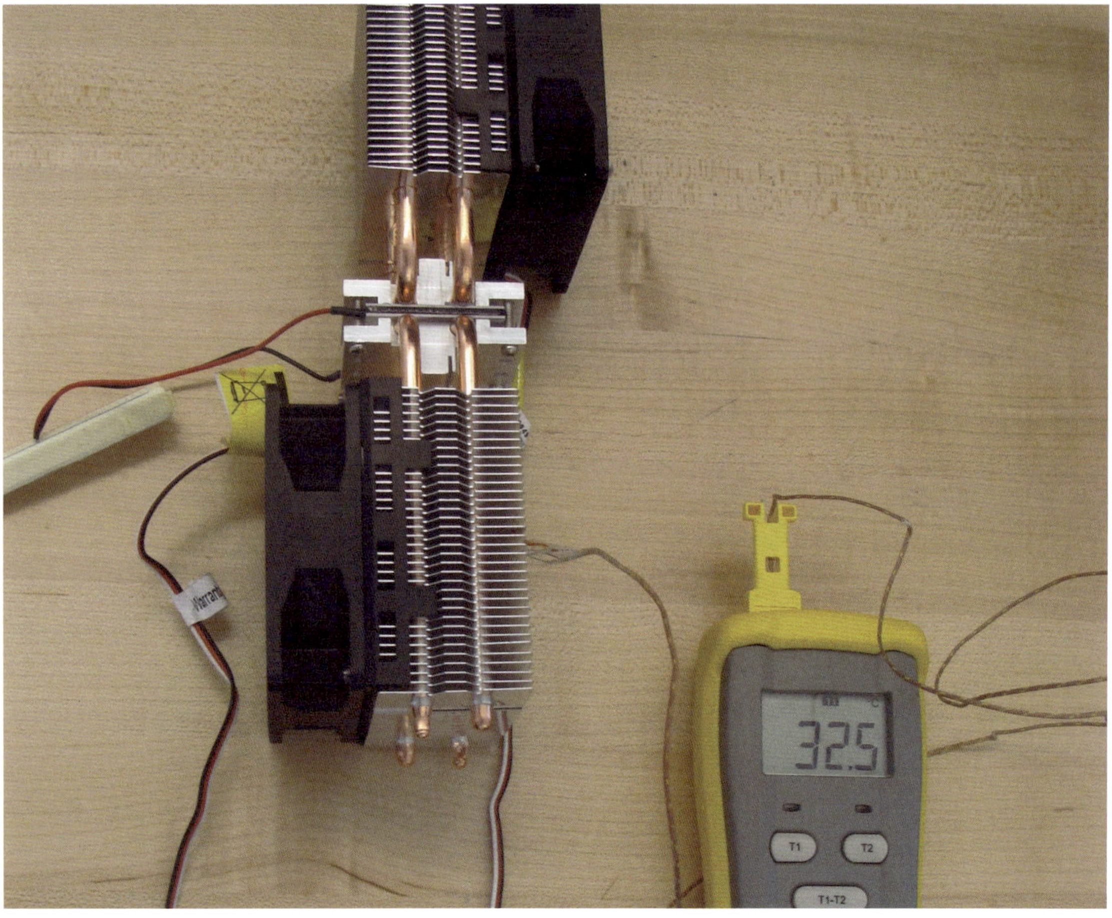
32.5
T1
T2
T1-T2

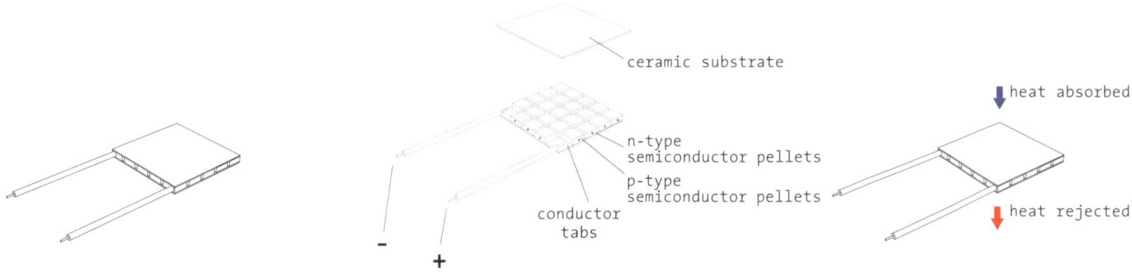

/06

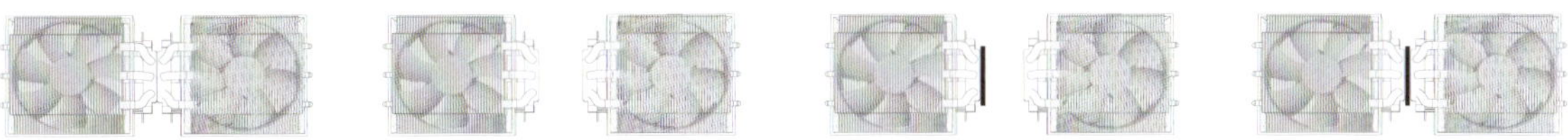

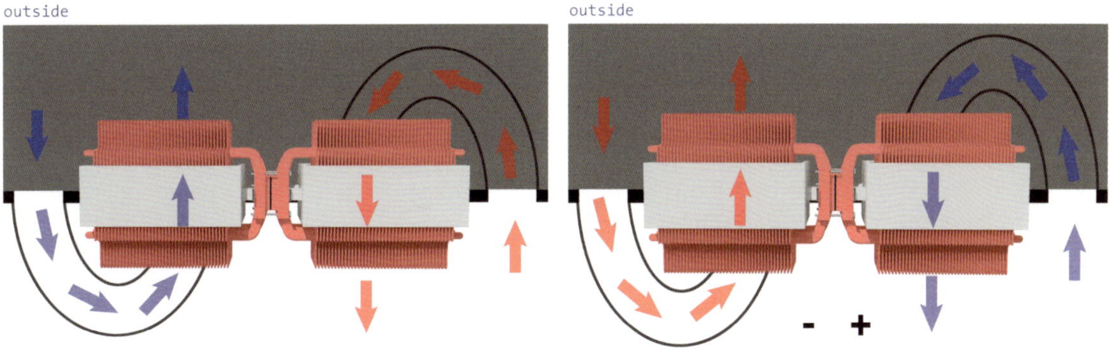

/06
The Peltier device is essentially a mini-heat pump; by reversing the current on the Peltier device, airflows can be adjusted according to the heating and cooling needs of the inhabitants.

variable geometry truss

A structural truss with a variable geometry that responds dynamically to eccentric loads / **David Long***

Variable Geometry Truss contributes to the growing field of Intelligent Responsive Architecture by using a novel approach to assembly, construction, modularity, and intelligence. It references 'Swarm' behaviour observed in nature in flocking birds, fish, and bees.

Responsive architecture combines built forms with integrated systems that are capable of responding to changing conditions. Responsive architecture is not merely concerned with physicality but is more focussed on behaviour. This emerging genre of architecture is at the forefront of an imminent paradigm shift in architectural design where buildings will no longer be static but rather dynamic objects which reflect the capabilities and requirements of the techno-environmentally conscious age.

/01
The design uses a swarm logic allowing each actuated truss member to continuously and independently adapt in response to local pressures measured and relayed by an array of sensors.

*Supervised by Rodolphe el-Khoury

/01

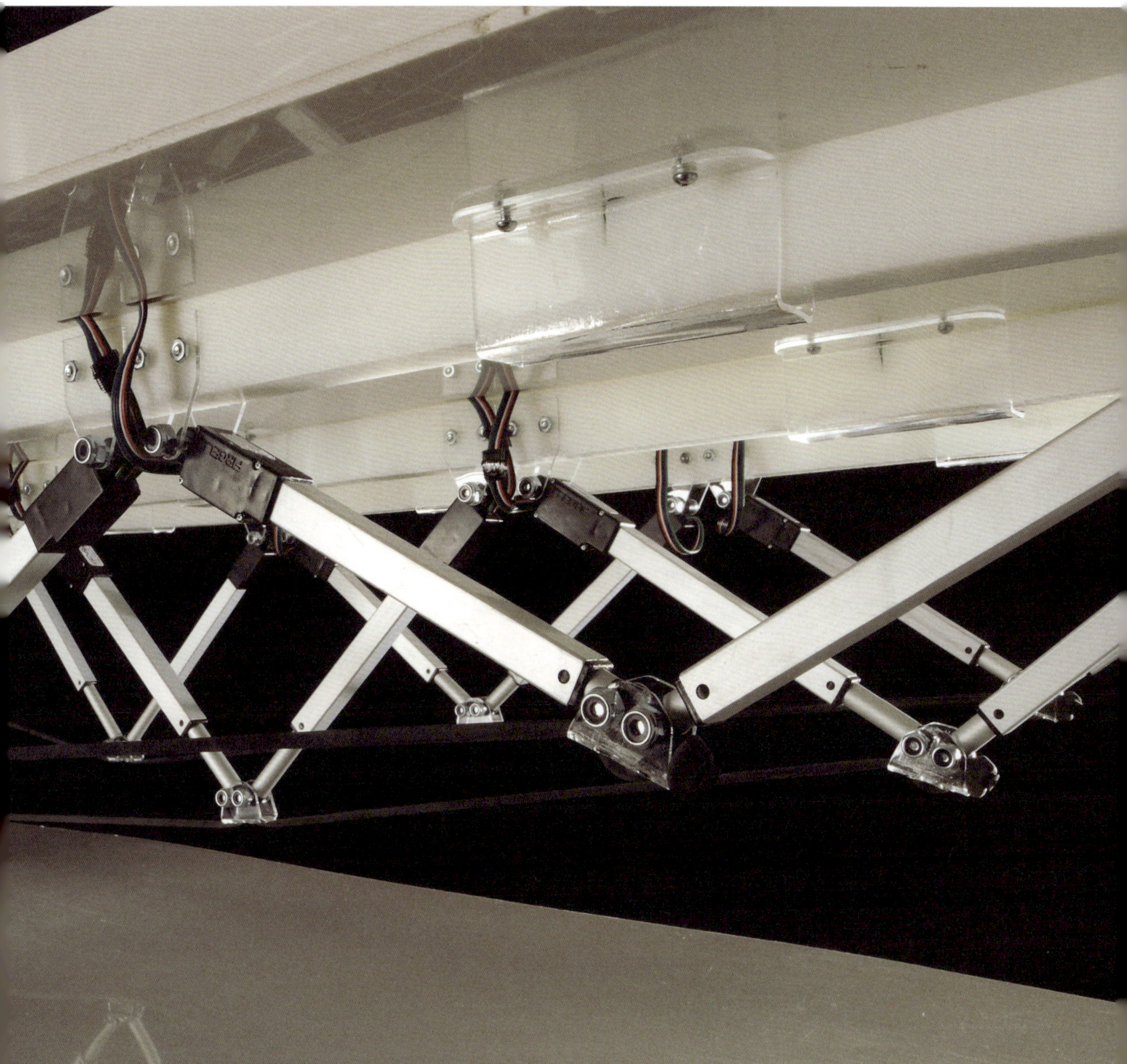

/02
Given a point load on the top chord, both webs (actuators) below that load should extend to the same length to distribute the load. To achieve this, the actuated responsive truss is implemented as a swarm; a single cell in the swarm consists of two actuators and a load sensor (positioned between the two actuators, on the top chord).

/03
Load is applied // Sensor detects compression // Actuators extend, decreasing compressive force // Actuators stop when force is sufficiently small.

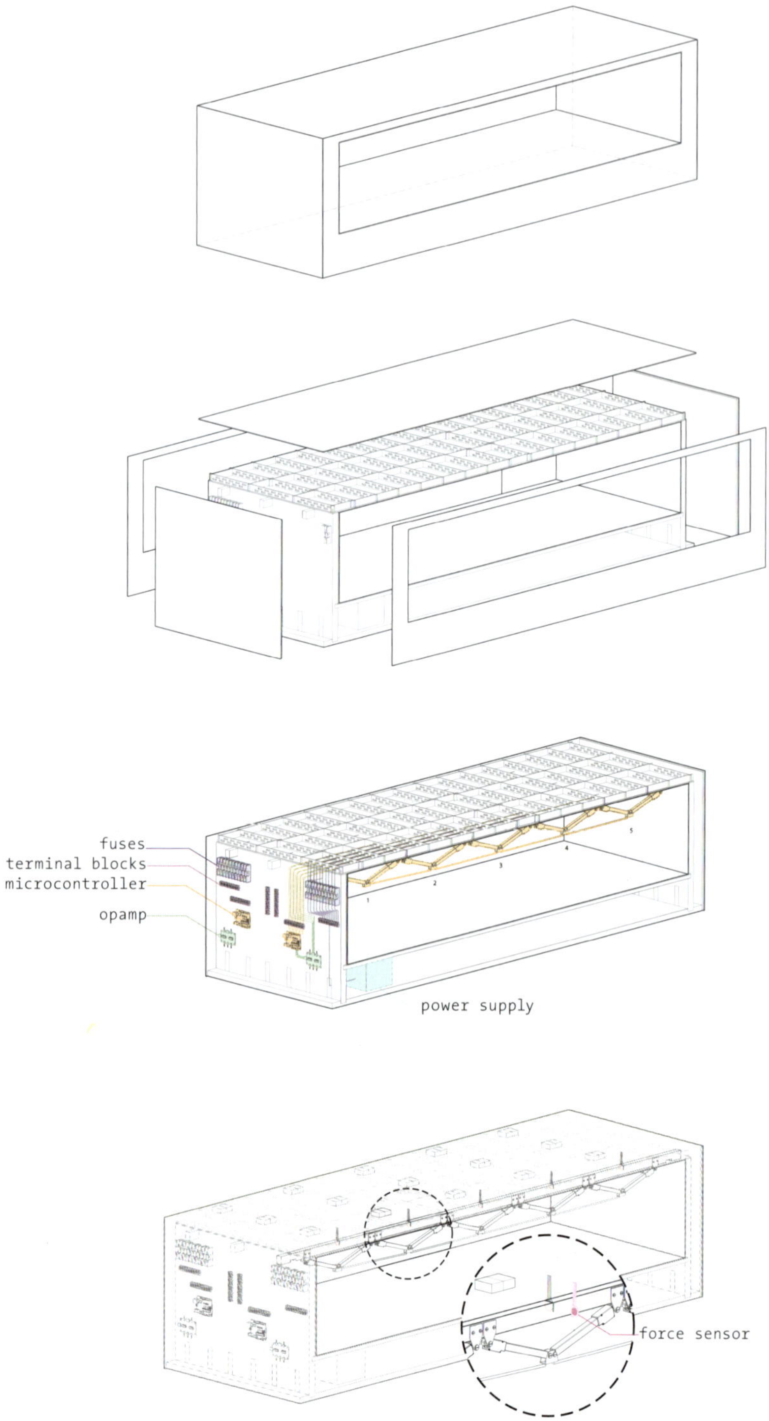

/02

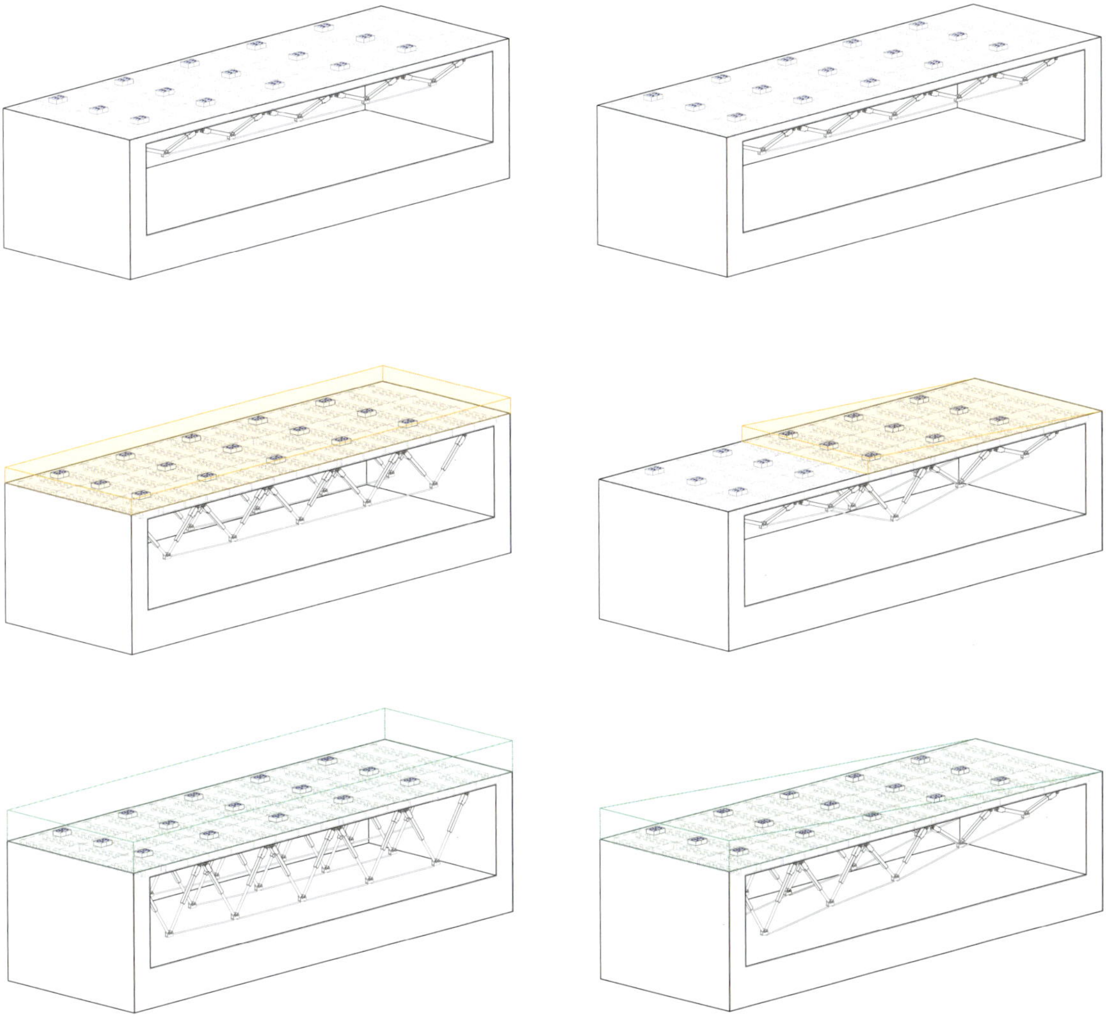

(a) load is applied

(b) sensor detects compression

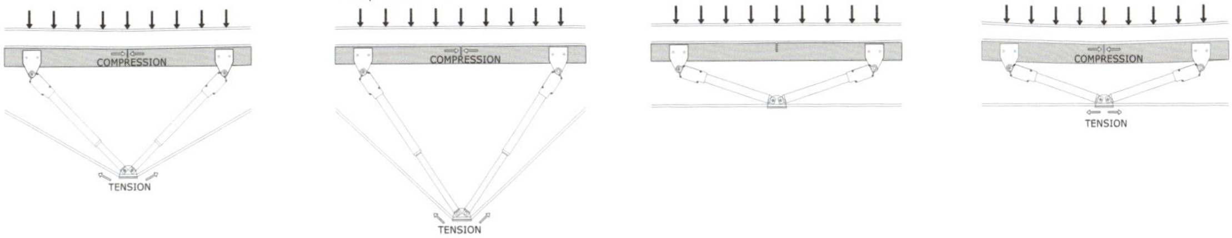

(c) actuators extend, decreasing compressive force.

(d) actuators stop when force is sufficiently small.

/03

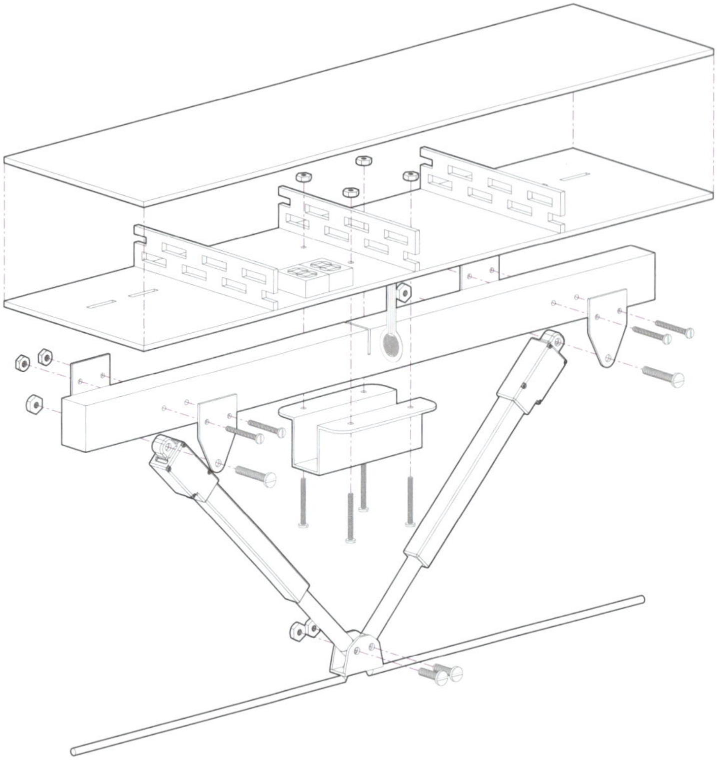

/04
Single cell detail.

/05
Dynamic prototype.

/06
On the left, convention centre with responsive actuated truss that optimises space, and on the right, the truss responds to accommodate maximum loads.

/04

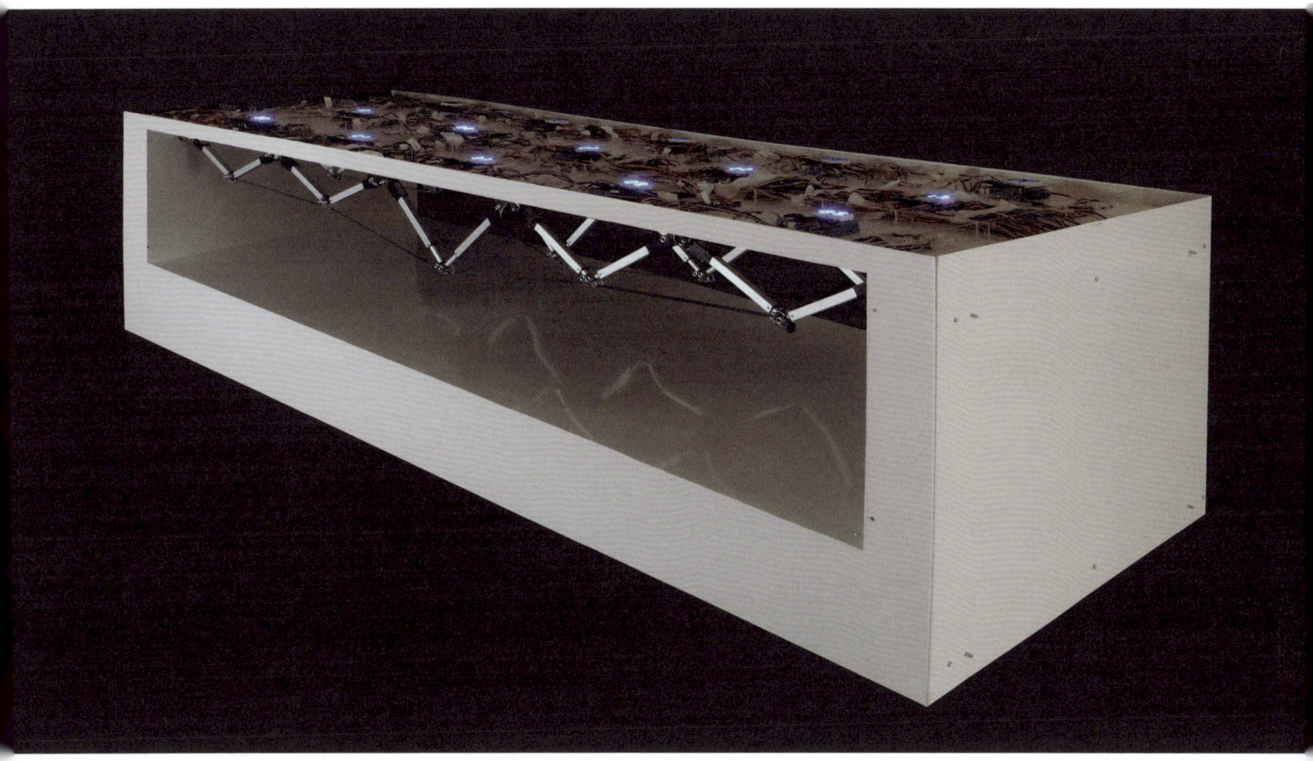

/05

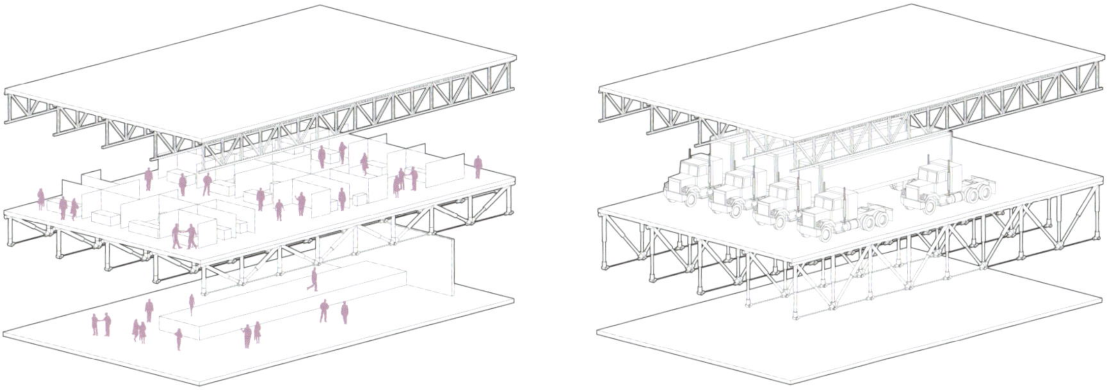

/06

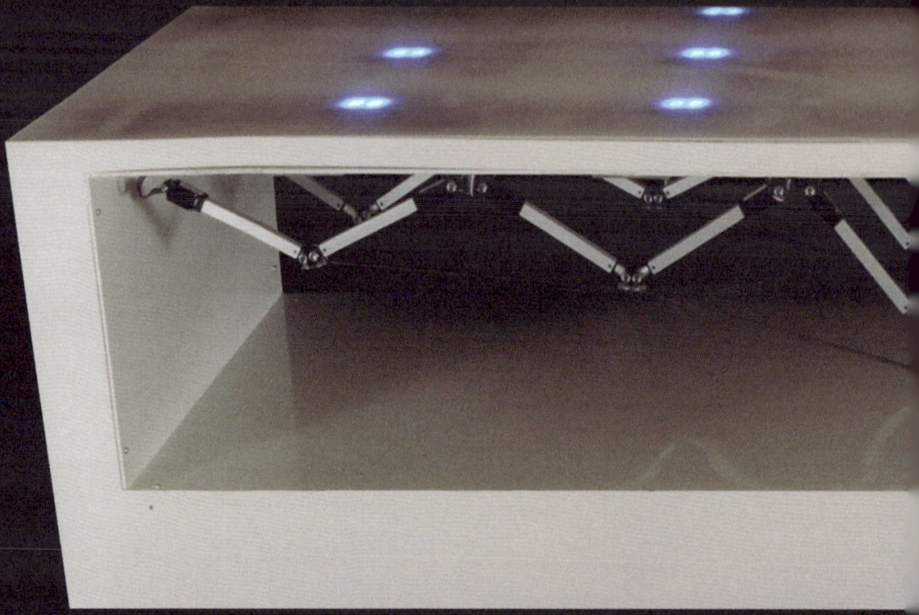

light-seeking brick

A brick wall is animated through responsive rotation based on temperature and lighting conditions / **Maggie Nelson***

Light-Seeking Bricks act as a kinetic cladding system with variable porosity to create variable aesthetic conditions within the home based on environmental readings.

Though bricks remain a fundamental building component worldwide, the brick wall is viewed as a fundamentally static object. Brickwork is rarely seen as an innovative material. By redesigning the brick unit to house responsive technologies, it is possible to animate a standard brick wall to create a dynamic system.

The Light-Seeking Bricks move through rotation. Their degree of rotation is conditioned based on temperature and exterior lighting conditions. By following an environmental logic, the brick system modulates heat and light gain through the façade for control of the interior space.

The kinetic bricks occur in alternating courses, sitting atop static bricks, within which are embedded light and temperature sensors that direct the brick rotation. Each section of static bricks is supported from the rear by a lightweight structure made of suspended cables or an alternative minimal system. The entire system acts as the rain screen portion of a double-skin façade, in front of a glass façade which functions as an air and water barrier and drainage plane.

Maintaining the brick form throughout the façade allows the interactive system to invoke the elements of surprise and wonder as a seemingly tradition material is operated in an extraordinary way.

/01
Prototype with kinetic bricks occurring in alternating courses atop static bricks. Light and temperature sensors to direct the brick rotation are embedded within each brick.

/02
This traditional building material can be animated Servo Motors.

*Guided by Rodolphe el-Khoury and Nashid Nabian

/01

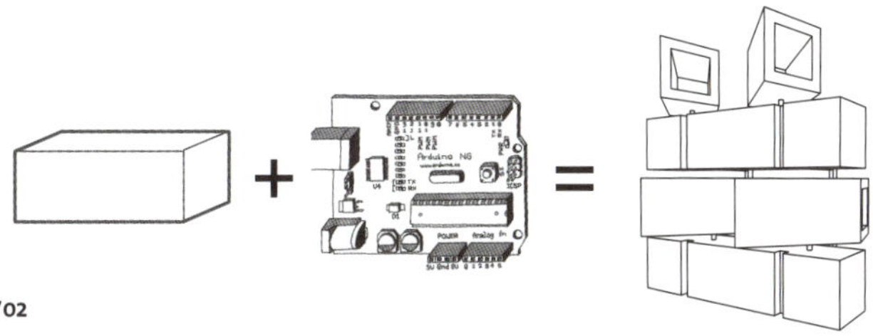

/02

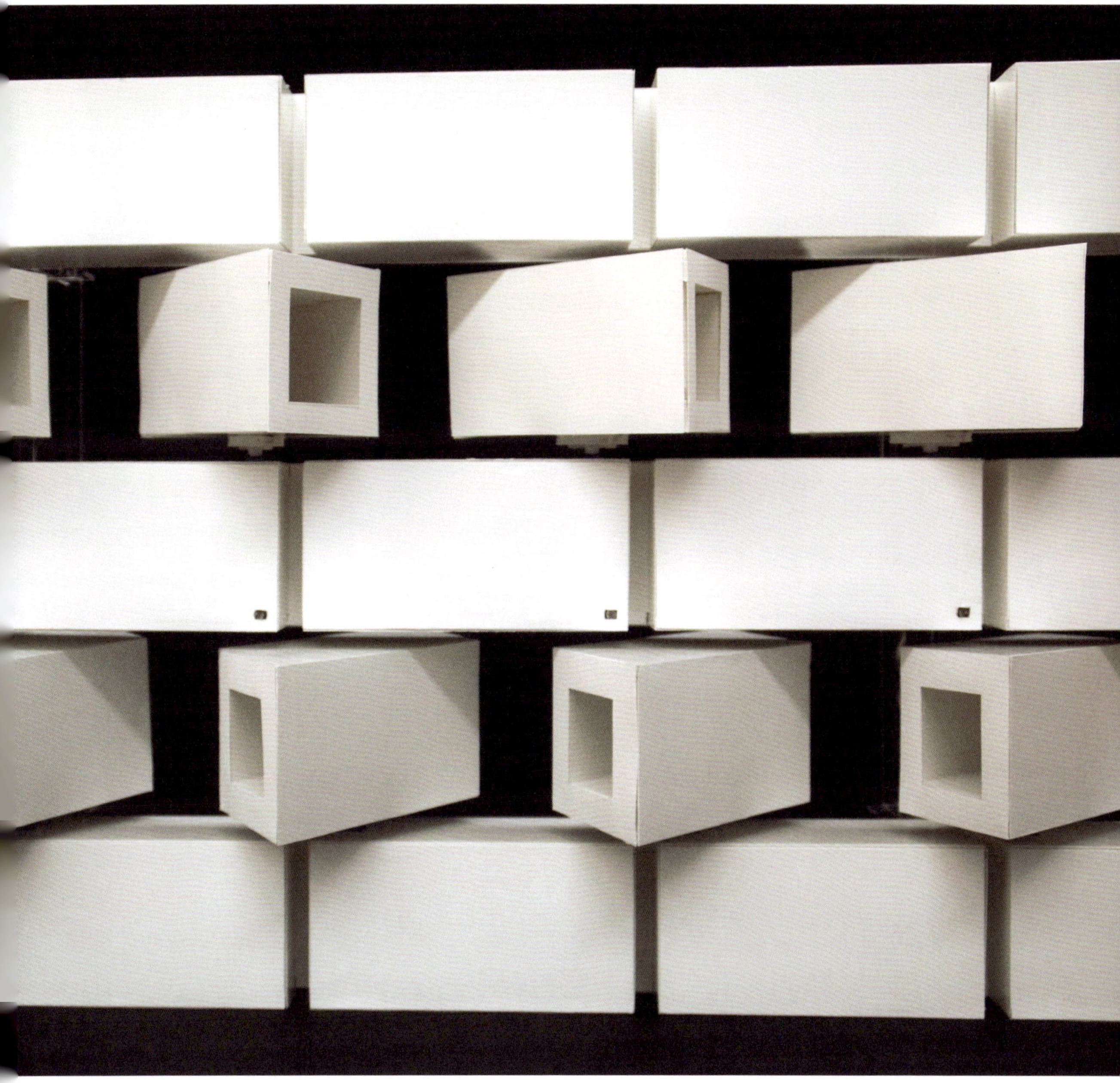

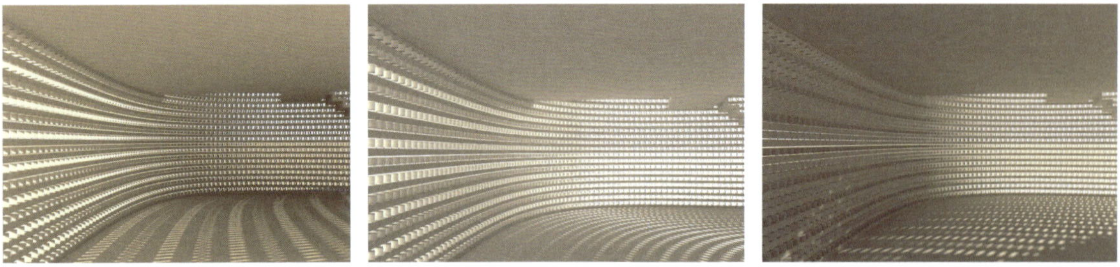

/03

/04

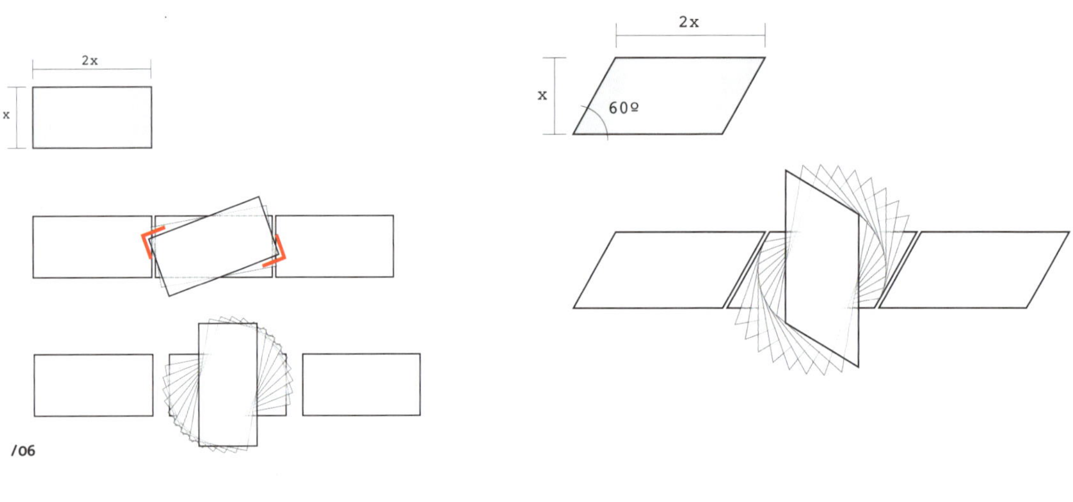

/06

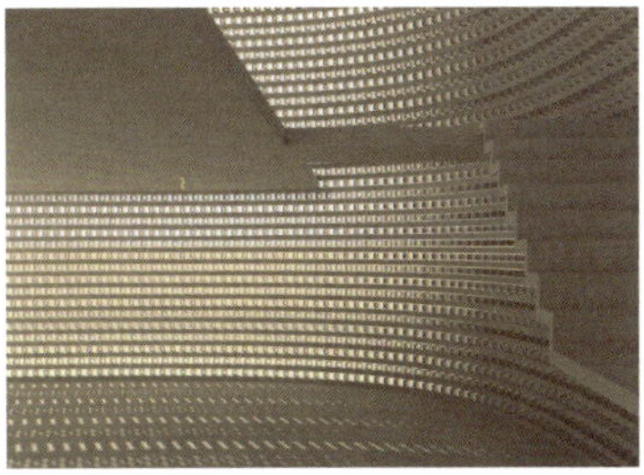

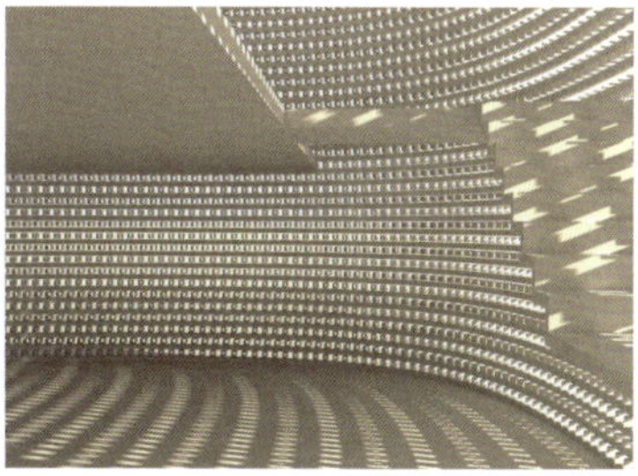

/05

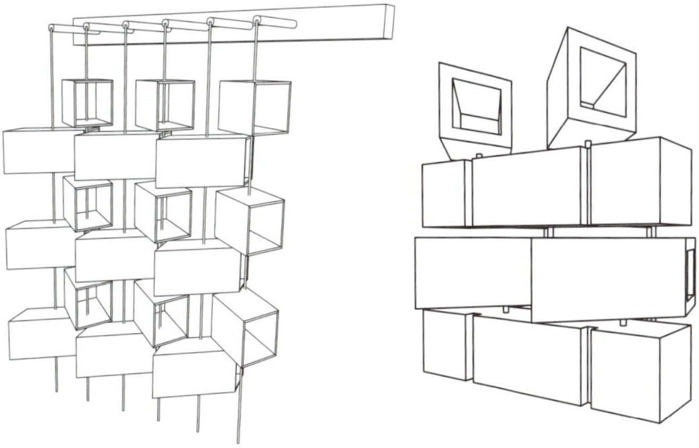

/07

/03
Renderings of rotating brick façade at various times of day. Responsive technologies.

/04
The rotating brick façade creates dappled light effects inside the home.

/05
Exterior view showing the array of brick applications at play in this sentient home.

/06
The shift from rectangular to parallelogram-shaped bricks facilitates a rotation of up to 90 degrees.

/07
The rotating brick system is supported by tension cables hung from a peripheral beam.

wallbot

Computerising the walls for novel spatial autonomy / **Otto Ng***

Wallbot is an intelligent wall unit, capable of reconfiguring itself in real time to respond to the behaviour or desire of the occupants.

Since Le Corbusier introduced the 'Domino' House in 1914, walls have been liberated from their historical role as loadbearing elements. Wallbot is a design investigation that attempts to even further liberate walls from their static past. The boundaries of space should be continuously shaped by the changing pressure of events and the unpredictable density of object flows, in addition to energy and comfort requirements.

Taking a typical loft in Boston as the principal site for the investigation, multiple configurations are achieved through a set of twenty-one metre wide wall sections. The transformation from one configuration to another requires translation, rotation and expansion of the wall pieces. In the interim, the wall sections will disperse, producing an exciting forest-like condition. Users communicate with Wallbots through an intuitive graphic interface or through direct gestural control. A spectrum of radically different spatial configurations can be produced to suit a spectrum of domestic conditions.

/01
Changing spatial conditions as Wallbots change configurations.

/02
Attaching the wall shell to a Wallbot prototype.

*Guided by Rodolphe el-Khoury and Nashid Nabian

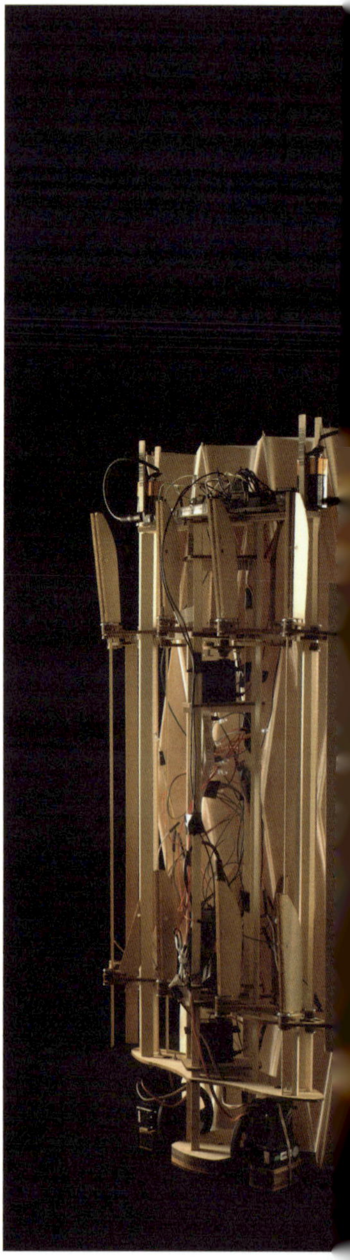

/01

/02

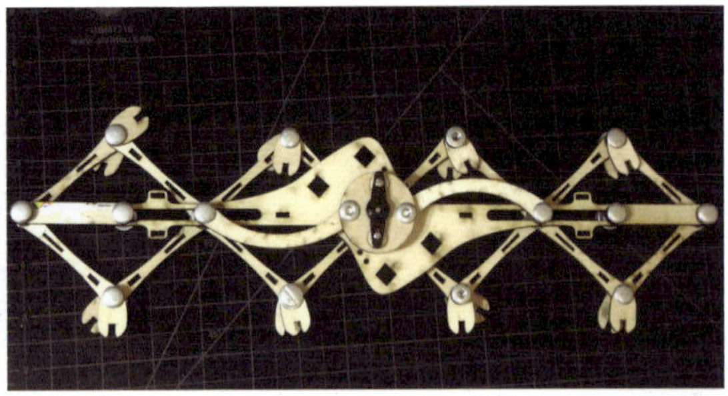

/03

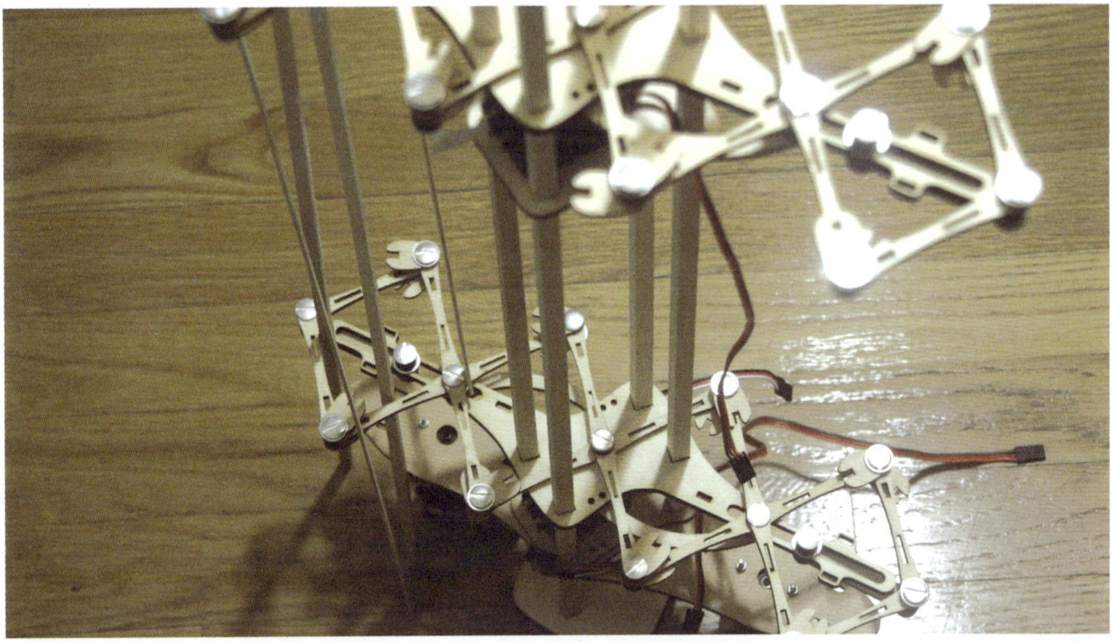

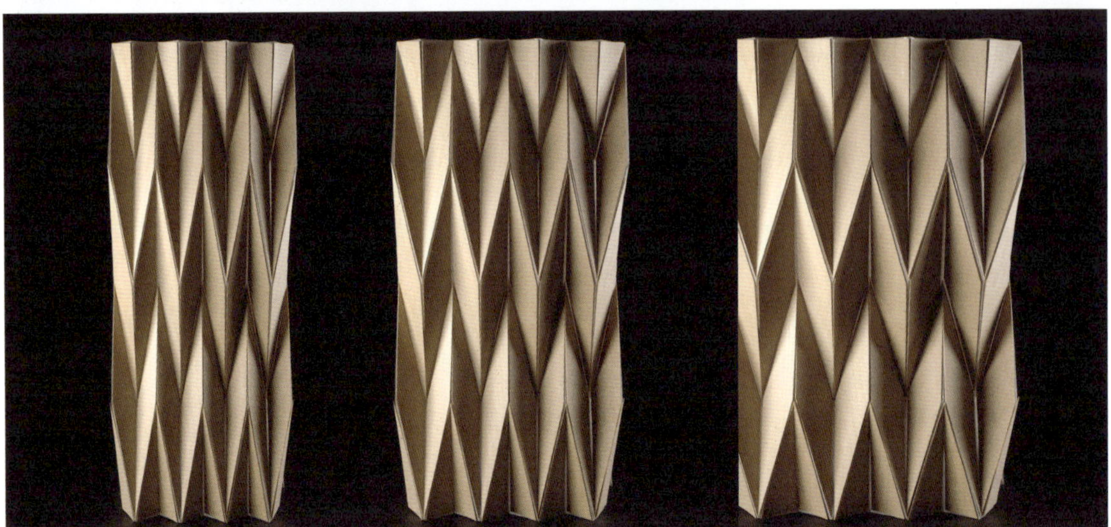

/04

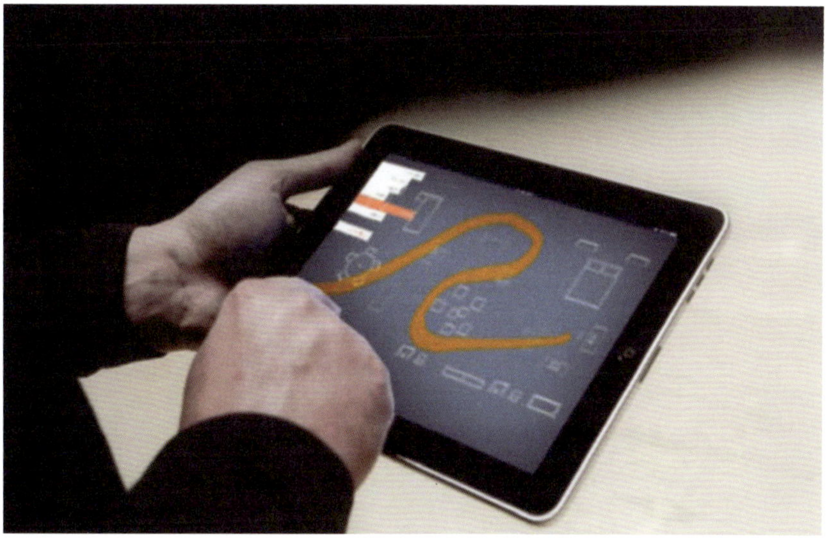

/05

/06

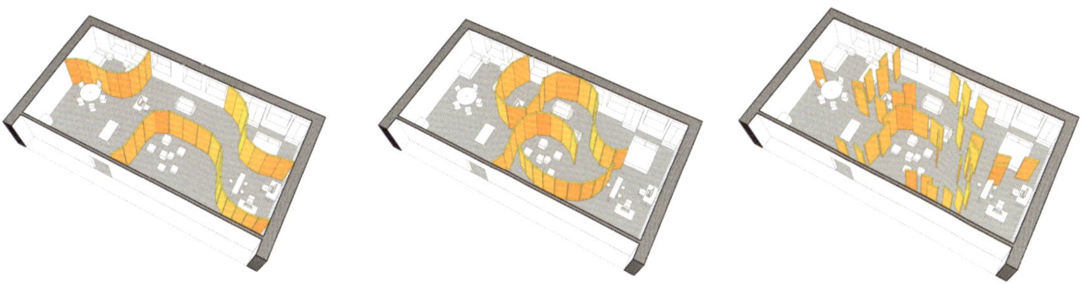

/07

/03
Prototype details.

/04
Expanding wall section: the shell material is rigid and flexible to allow expansion. Pleating allows the rigid shell to transform through expansion and contraction.

/05
The user can communicate with Wallbots through an intuitive graphic interface or through direct gestural control.

/06
Swarm condition of prototypes.

/07
Transition between two configurations.

variable cladding

A building's capacity to respond and adapt to varying conditions is most useful at the threshold between the interior and exterior. Whereas traditional structure favoured an airtight boundary, responsive systems enable permeable envelopes that have a capacity to productively negotiate rather than simply prevent contact between varying extremes. The following projects explore scenarios and technologies that fashion cladding elements like epidermal cells with a capacity to intelligently modulate the dynamic exchanges of energy, light, air, and water through a sensing and responsive skin.

swarm tile

Interactive Modular Tile System
/ **Shadi Ramos***

Swarm Tile is an interactive modular tiling system. The tiles react to their surrounding environment through sensing technology, or respond to more global conditions of their context of use and the occupancy patterns of the tiled space through network connections among all the tiles.

With a very simple and straightforward architecture, the proposed electronically enhanced, responsive system of Swarm Tiles examines the spatial and conceptual ramifications of a very interesting concept: since so much of the physical world is based on modular units of materials, what would happen if these modules possessed some of the attributes of the biological cell network that constitutes a living creature? What if they were able to sense their context and augment their surrounding environment with programmable behaviours?

Each cell or module is composed of a sensor, multiple light emitting diodes (LEDs), wireless communication, and an embedded micro-controller. For prototyping purposes, a simple push button registers that the tile is stepped on and sends this information to an Arduino board. In one scenario, the board activates the local condition by regulating embedded LEDs in response to the registered context. In another scenario, the locally registered condition is wirelessly communicated with other tiles via a mounted ZIGBEE radio. This allows for the creation of a network condition among the tiles where each responds to a collectively evaluated context based on cellular automata logic: if a stimulus is present, the information is processed and an appropriate response is sent back to the cell itself and to other cells in the system.

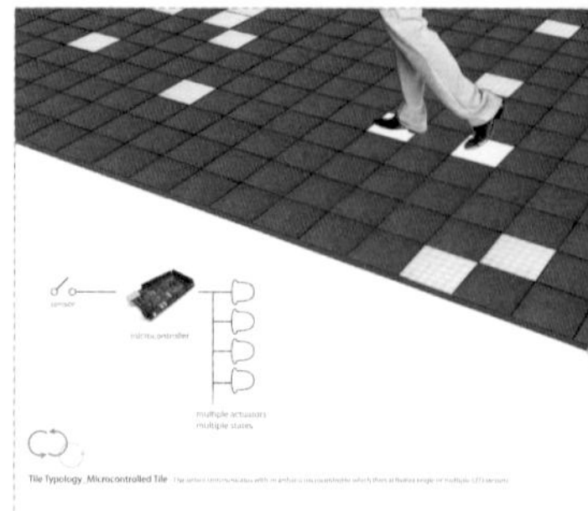

/01

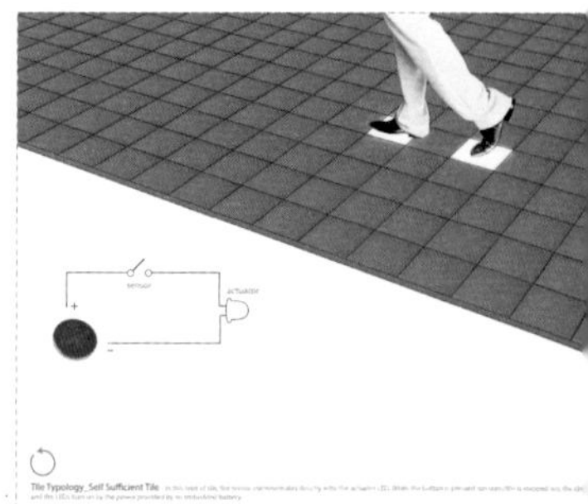

/02

/01
Microcontrolled Tile. The sensor communicates with an Arduino microcontroller which then activates single or multiple LED sensors.

/02
Self-sufficient Tile. The sensor communicates directly with the actuator LED. When the button is pressed (tile is stepped on), the circuit is completed and the LEDs turn on.

/03
Super Tile. This model operates in a similar manner to the microcontrolled tile. However, this tile is capable of communicating with other tiles or a computer through Xbee, a wireless network adaptor. This allows for virtual collection and analysis of data from a remote source. The data can be fed to the processing software and visualised for better usage.

*Guided by Rodolphe el-Khoury and Nashid Nabian

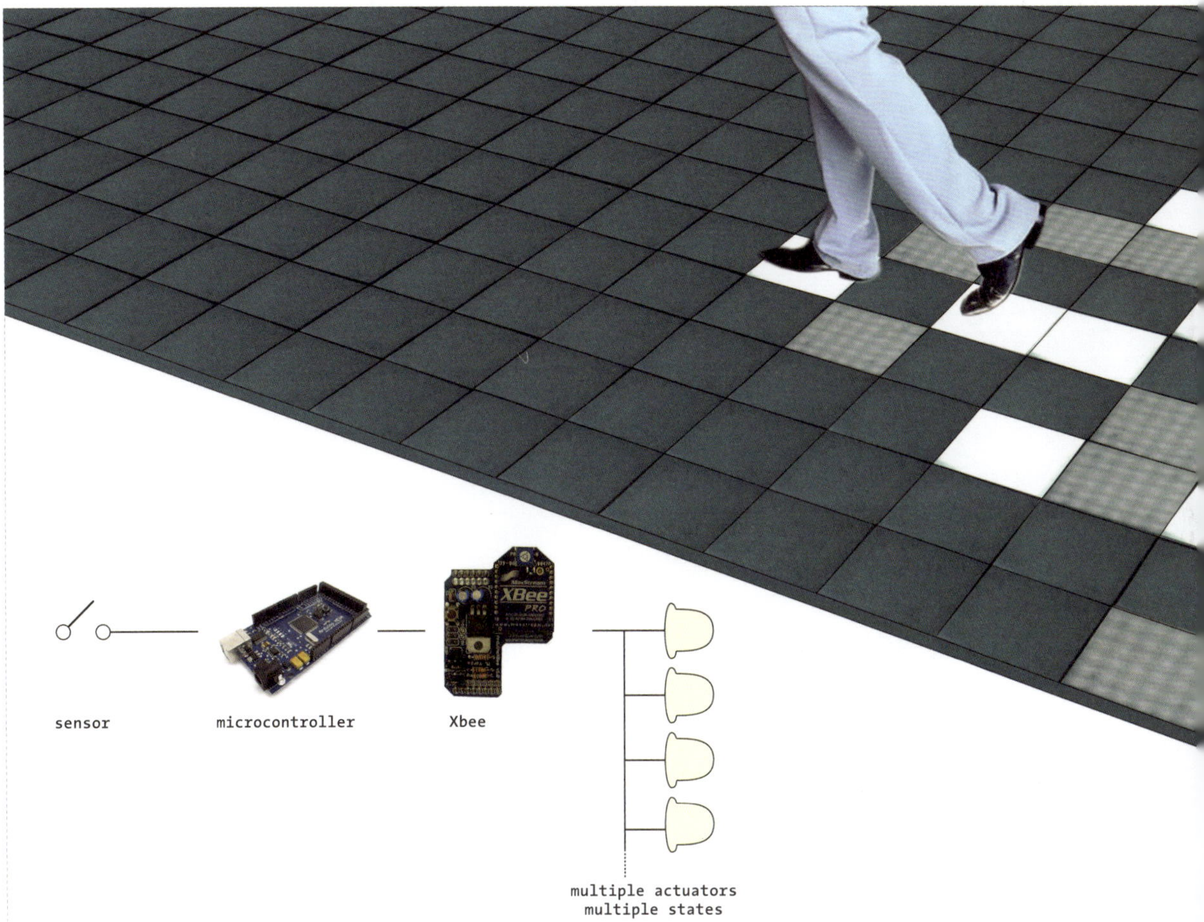
sensor
microcontroller
Xbee
multiple actuators
multiple states

/03

/04

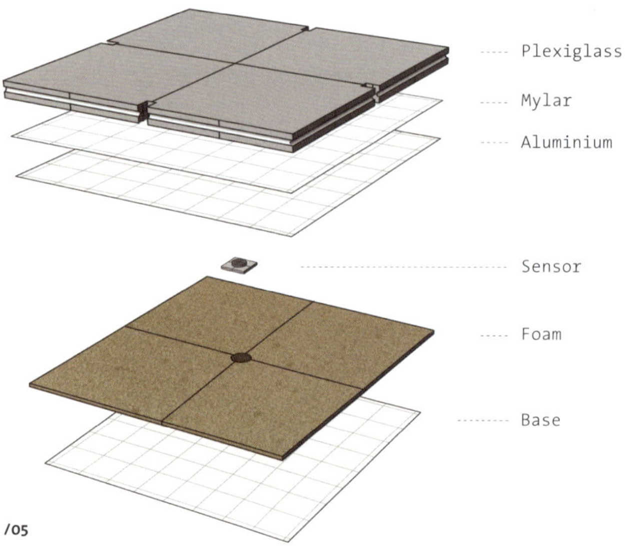

/05

/04
Swarm Tile demonstration.

/05
Swarm Tile components showing material support, light diffusing elements and a pressure sensing mechanism.

/06
Super Tile detail. Each tile in grid A activates itself and its equivalent tile in grid B.

/07
Construction specifications for the tiles.

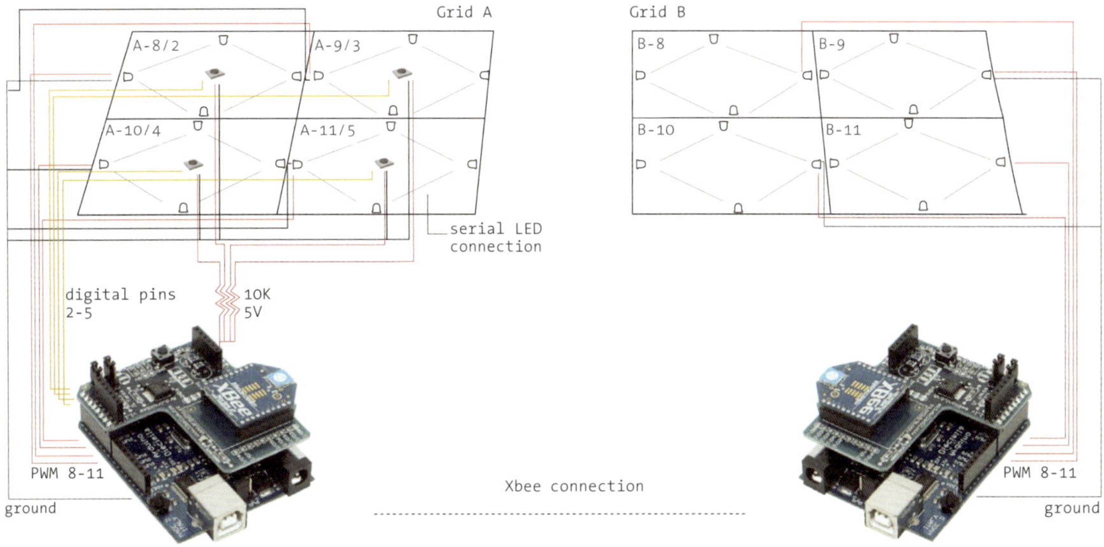

Grid A
A-8/2
A-9/3
A-10/4
A-11/5
serial LED
connection
digital pins
2-5
10K
5V
PWM 8-11
ground
Grid B
B-8
B-9
B-10
B-11
Xbee connection
PWM 8-11
ground

/06

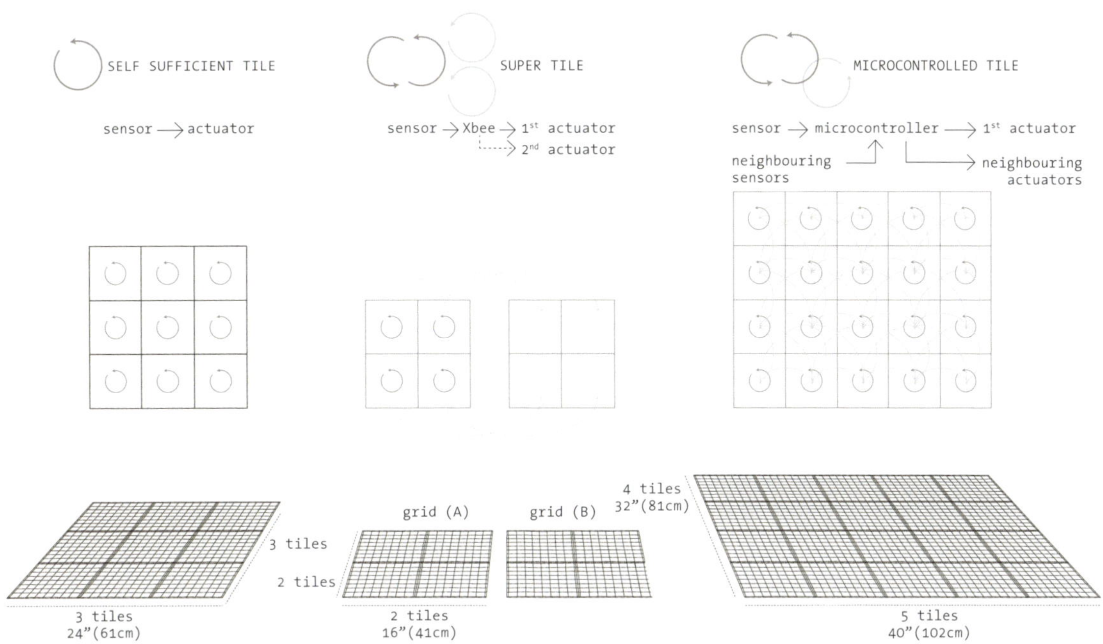

SELF SUFFICIENT TILE
sensor → actuator
SUPER TILE
sensor → Xbee → 1st actuator
2nd actuator
MICROCONTROLLED TILE
sensor → microcontroller → 1st actuator
neighbouring
sensors
neighbouring
actuators
3 tiles
3 tiles
24"(61cm)
grid (A)
grid (B)
2 tiles
2 tiles
16"(41cm)
4 tiles
32"(81cm)
5 tiles
40"(102cm)

/07

pneumatic envelope

A thermally responsive insulating curtain wall with variable transparency / **Rick Sole***

The Pneumatic Wall is a thermally responsive insulating curtain wall with variable transparency, oscillating in the gradient between wall and window.

The Pneumatic Wall has a cellular structure, with each cell or air-pocket consisting of three membranes forming two separately controlled pneumatic pockets. Through sensor data processed in heat transfer analysis software, individual pneumatic pockets are controlled to inflate or deflate as required by user preference, or by changing climatic conditions.

Passive convective air flows are created based on thermal gradients on the interior side. By dynamically altering the thermal properties of a wall in section, a variety of air circulation patterns are achieved within the space. Operable windows take the form of air pockets on a central revolving axis which can be rotated to provide fresh air. Each membrane has its own pattern of opacity and transparency. By modulating the air pockets, different transparencies are achieved. The curtain wall changes in appearance based on the selected interior condition, or level of insulation.

/01

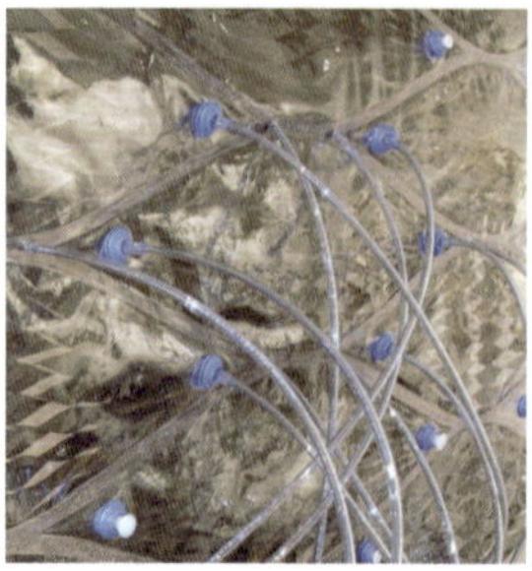

/01
The Pneumatic Envelope fully inflated. The more inflated, the more insulating, and the more transparent the wall becomes.

*Guided by Christos Marcopoulos and Carol Moukheiber

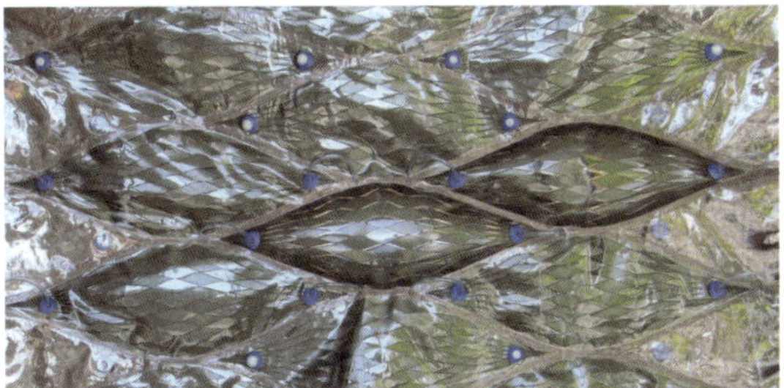

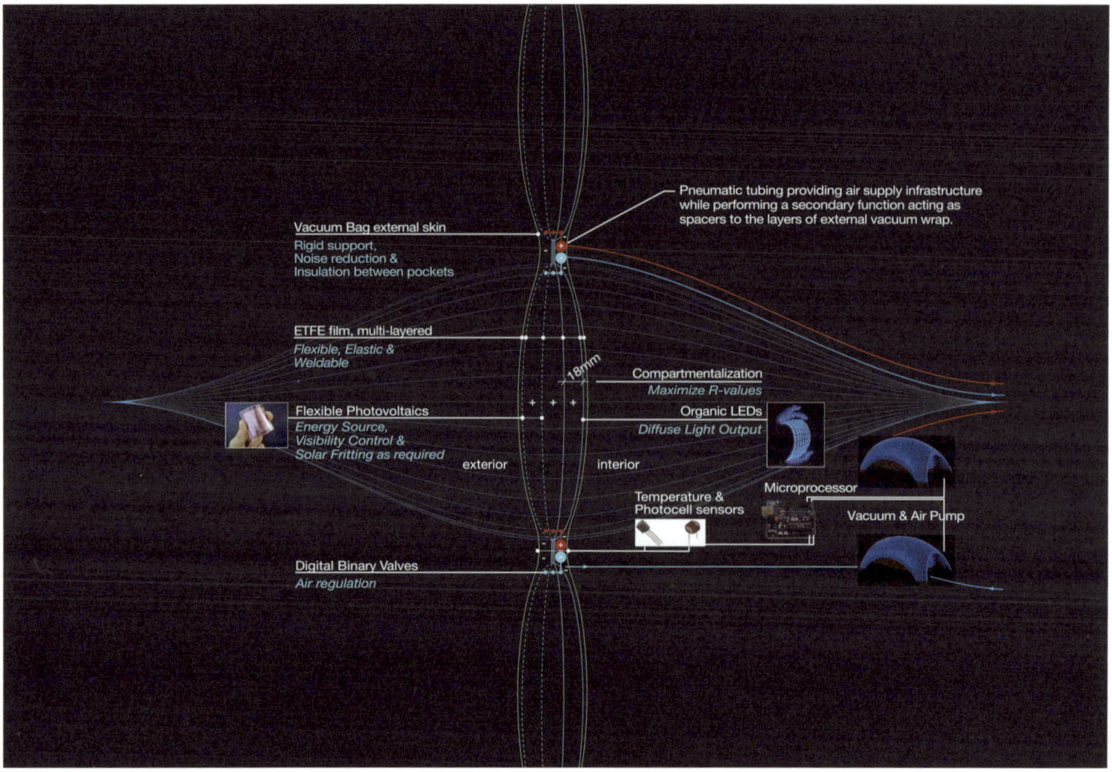
Pneumatic tubing providing air supply infrastructure while performing a secondary function acting as spacers to the layers of external vacuum wrap.
Vacuum Bag external skin
Rigid support, Noise reduction & Insulation between pockets
ETFE film, multi-layered
Flexible, Elastic & Weldable
18mm
Compartmentalization
Maximize R-values
Flexible Photovoltaics
Energy Source, Visibility Control & Solar Fritting as required
Organic LEDs
Diffuse Light Output
exterior
interior
Microprocessor
Temperature & Photocell sensors
Vacuum & Air Pump
Digital Binary Valves
Air regulation

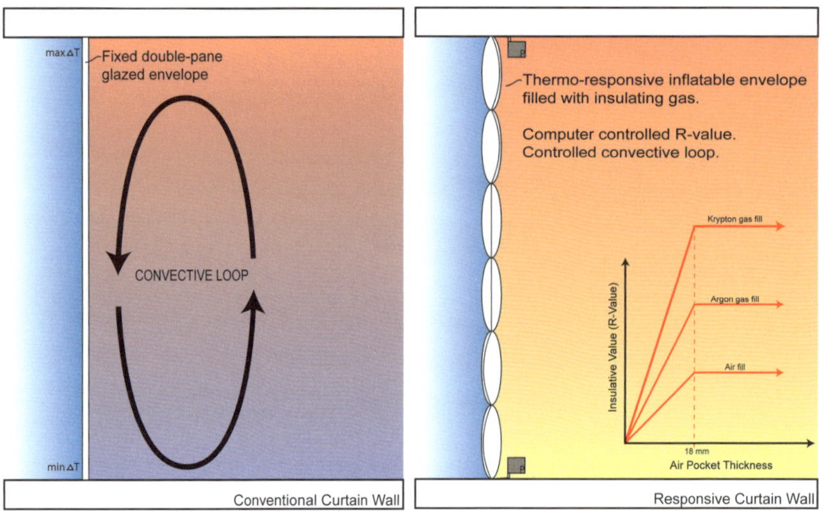
max∆T
Fixed double-pane glazed envelope
CONVECTIVE LOOP
min∆T
Conventional Curtain Wall
Thermo-responsive inflatable envelope filled with insulating gas.
Computer controlled R-value.
Controlled convective loop.
Krypton gas fill
Argon gas fill
Air fill
Insulative Value (R-Value)
18 mm
Air Pocket Thickness
Responsive Curtain Wall

/02

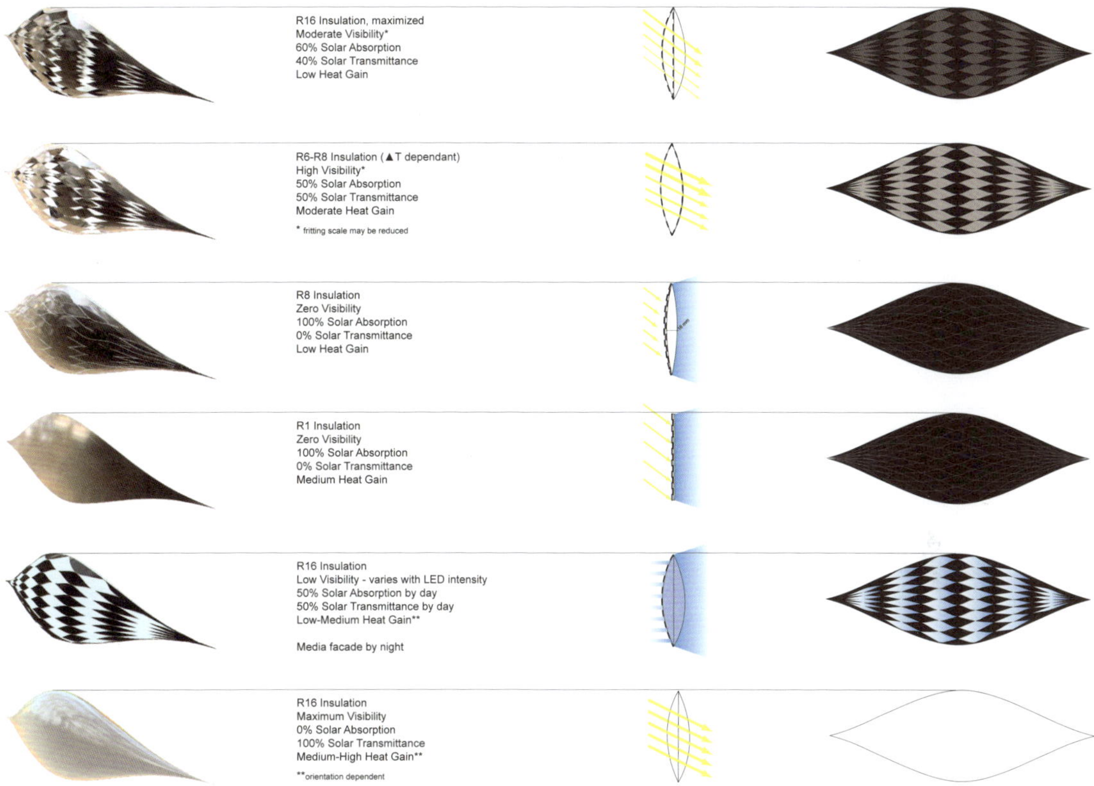

/03

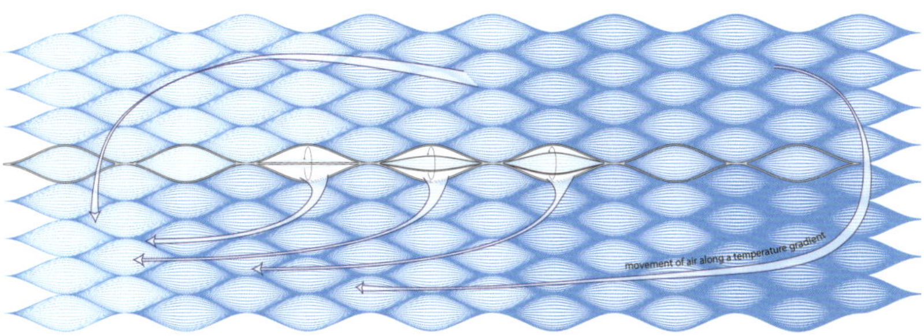

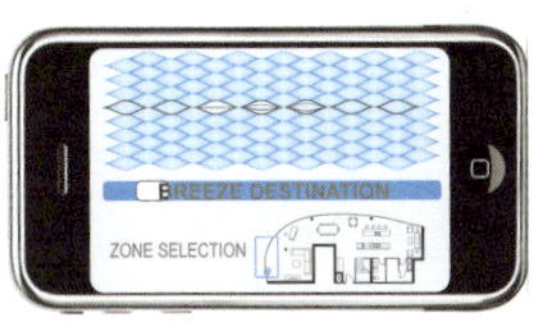

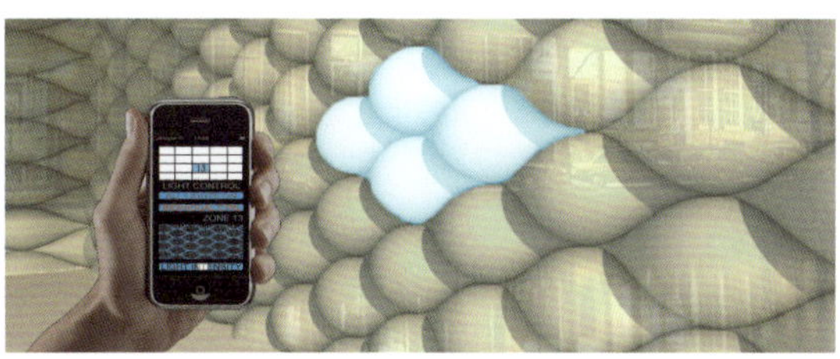

/04

/02
Each 'pillow' inflates or deflates as necessary to provide local thermal comfort. A variation in insulation levels across the wall's vertical section allows for the generation of convective flows within a space, alleviating the problem of vertical heat entrapment.

/03
Users can speak to the wall remotely and activate its thermal properties at will.

/04
Operable cells pivot along their central axis.

poly-glazed curtain wall

A multi-layered glazing system that harvests energy and sets up interior mood / **Ali Yarbakhti***

The Poly-Glazed Curtain Wall is an interactive cladding system that responds intelligently to changing external circumstances and internal performance requirements. Each panel is capable of projecting the diversity inside the building to the outside while maintaining optimal visual and thermal conditions indoors.

The proposed electronically enhanced architecture of the Poly-Glazed Curtain Wall cladding system consists of multi-layer striped glass sheets that transform the architectural façade to a dynamic mechanism capable of changing colour and transparency in response to interior and exterior circumstances. Since the proposal includes the deployment of a transparent solar panel as an embedded layer, the panel is self-sufficient in terms of energy consumption.

Each layer in the panel consists of alternating transparent and coloured translucent strips, and is placed on a geared shaft connected to a micro servo motor. By changing the rotational degree of each servo depending on the pattern in which the sliding layers are placed in relation to each other, the coloured translucent strips are juxtaposed to create a colour composition that is unique to a particular setting. For prototyping purposes, each panel has an embedded Arduino board that dynamically changes the arrangement of layers by controlling the servo angles. The context is registered via a presence sensor that identifies the level of activity inside. The users also can override the automated colour setting via direct input. In other possible implementation scenarios, the panels can respond to temperature or light intensity and apply algorithms that computationally decide on the optimum transparency/colour combination in response to changing environmental circumstances.

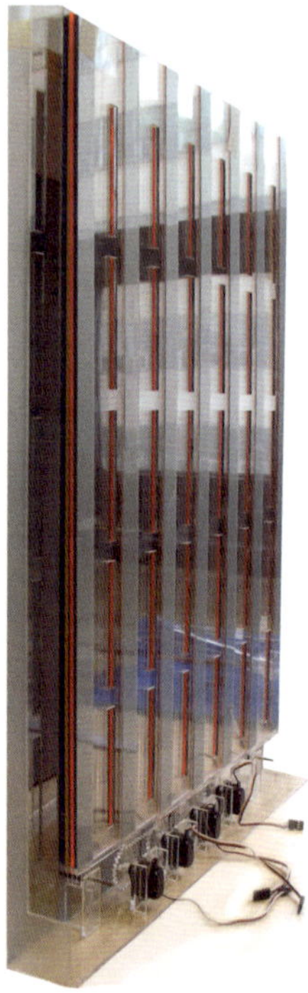

/01

/01
Detail of layered glazing and chromatic filtering.

*Guided by Rodolphe el-Khoury and Nashid Nabian

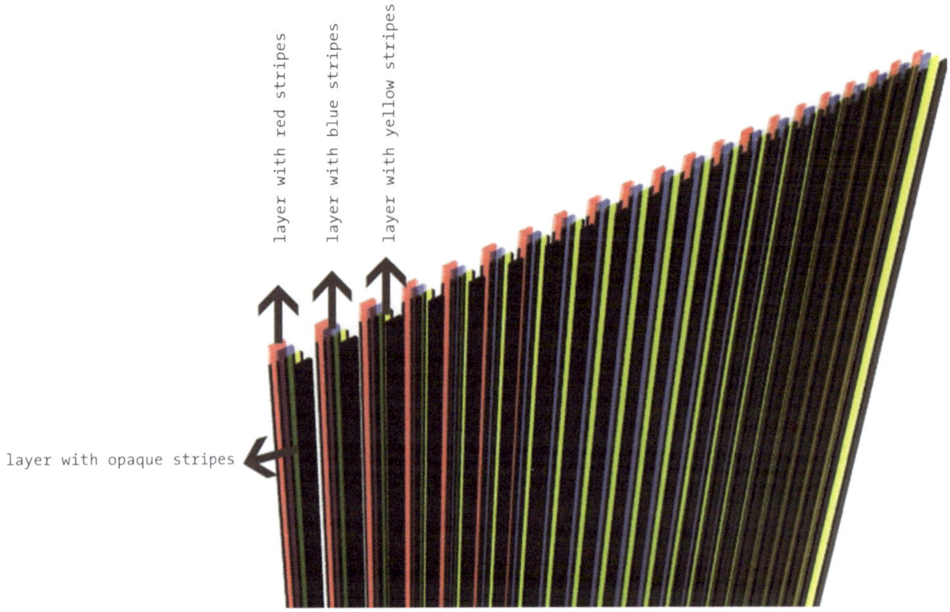
layer with red stripes
layer with blue stripes
layer with yellow stripes
layer with opaque stripes

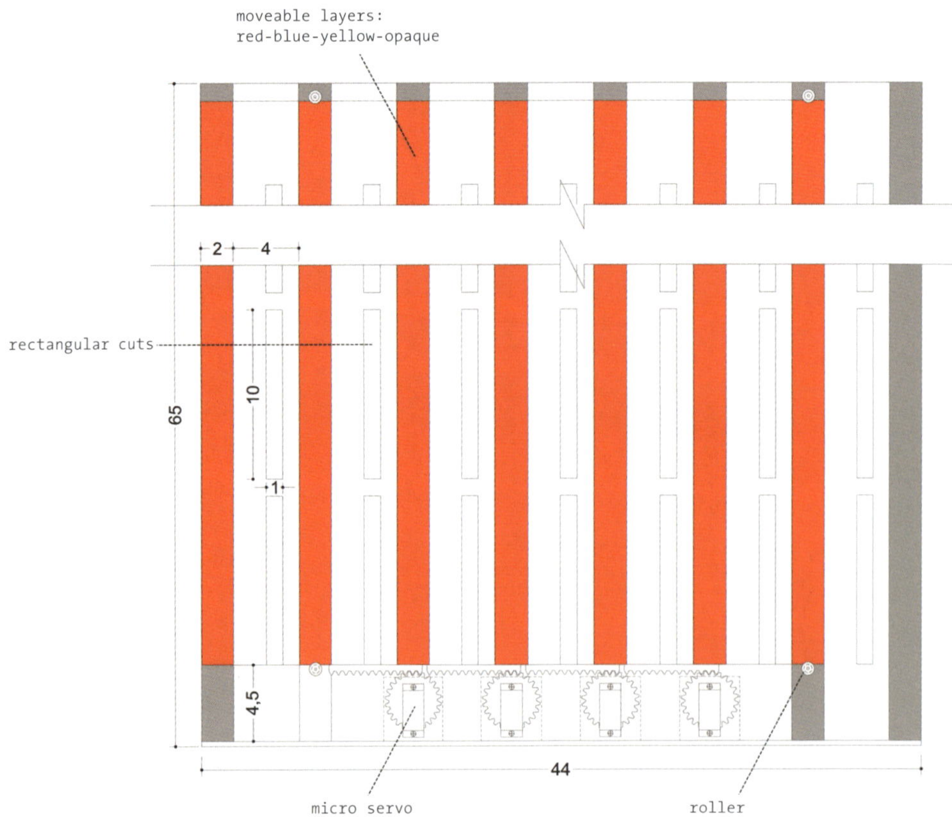
moveable layers:
red-blue-yellow-opaque
2
4
rectangular cuts
10
65
1
4,5
44
micro servo
roller

/03

/02

Diagrams detailing multi-layered glazing system and mechanism of servo-activated sliding.

/03

Various scenarios of light-filtering and mood-setting configurations in anticipation of or in response to different activities.

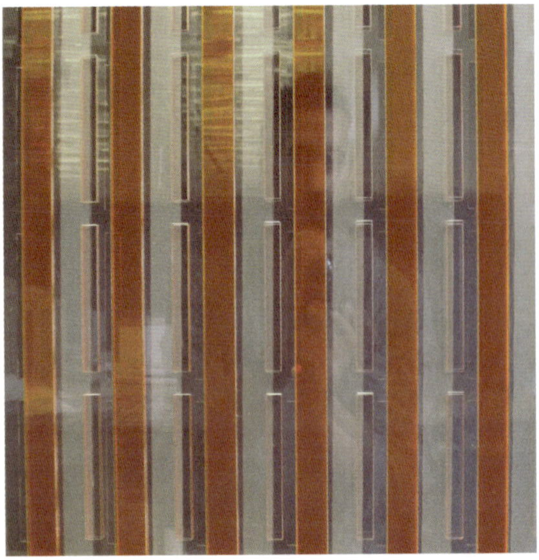

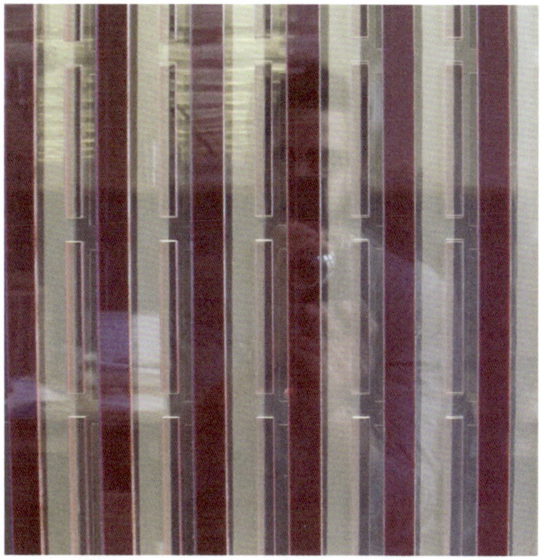

iviron microclimate

Individual environments are customised to respond to their inhabitants
/ **Julianna Sassaman***

Iviron forms a series of customisable microclimates, created when inhabitants pass between arrays of responsive floor and ceiling tiles.

Iviron uses haptic actuators, including lights, heaters and speakers, which are activated as the inhabitant passes through the array. Sensors in the Iviron tiles trigger the activation of these embedded technologies. The actuators are configured to read a personal identity profile carried by the inhabitant, who communicates his or her climate preferences.

Iviron tiles vary in size, depending on the nature of the particular actuator embedded within them, and on the actuator's sphere of influence.

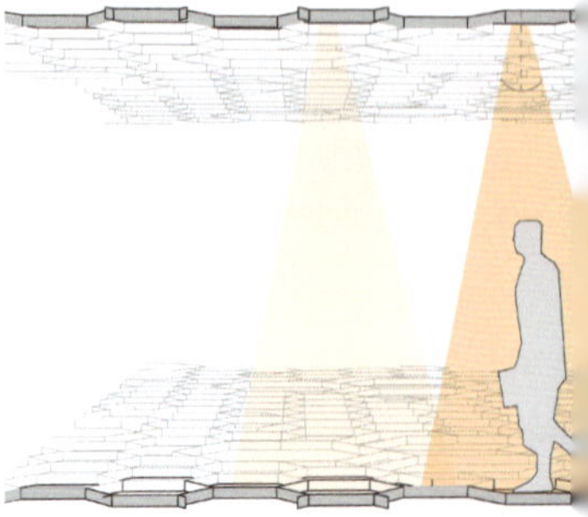

/01

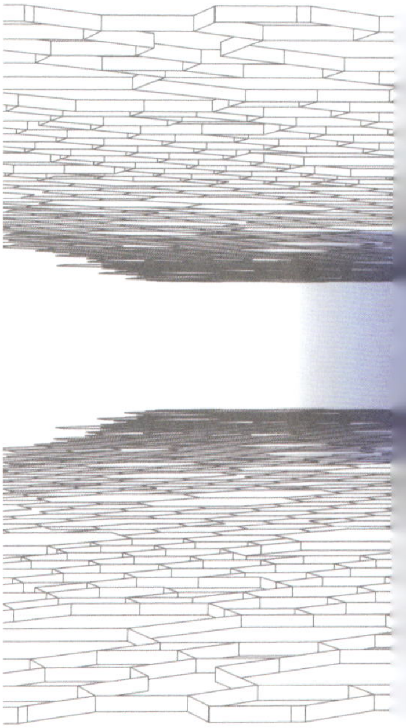

/02

/01
Embedded technologies include lighting, sound and heat.

/02
The result is a dynamic open space.

*Guided by Rodolphe el-Khoury and Nashid Nabian

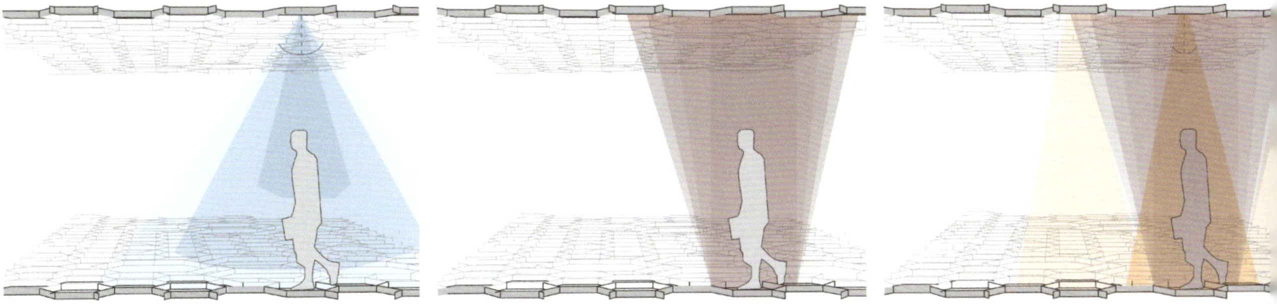

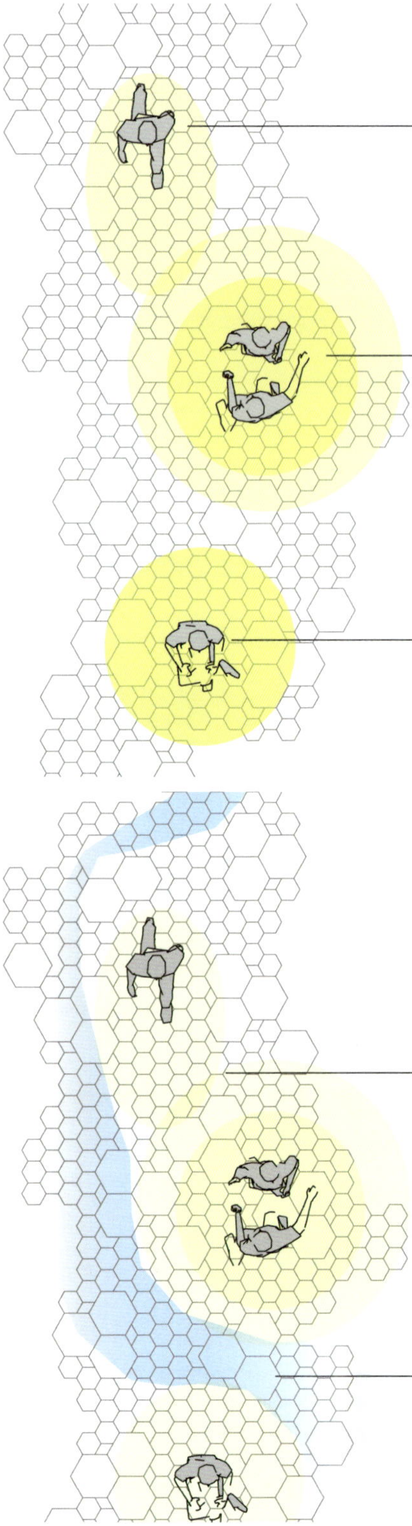

fast moving person

the microclimate of a moving person will vaguely anticipate their action, while submitting other more stable microclimates

group of people

when people gather, their microclimate is additive and activates a much larger area than either individual would

stationary person

this person's microclimate will hold steady even with the disturbance of a passing person

overlap zones

here microclimates must act according to the hierarchy of space - in this case time occupation: the longer user environment trumps a faster environment

dissipation zones

embedded micro fans keep the unused spaces working as low level HVAC and as a disturbance to micro environment tails

/03

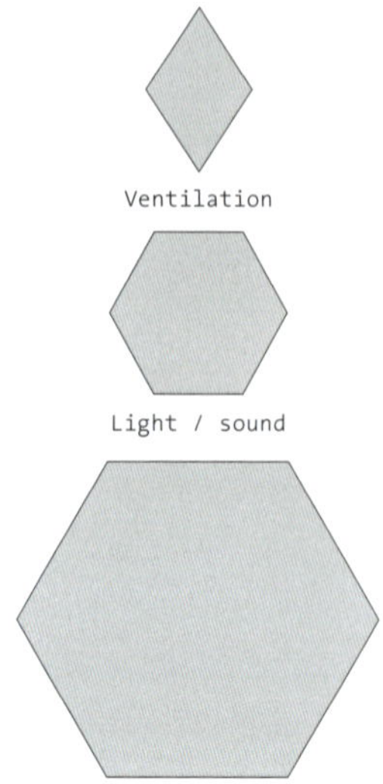

/04

/05

/06

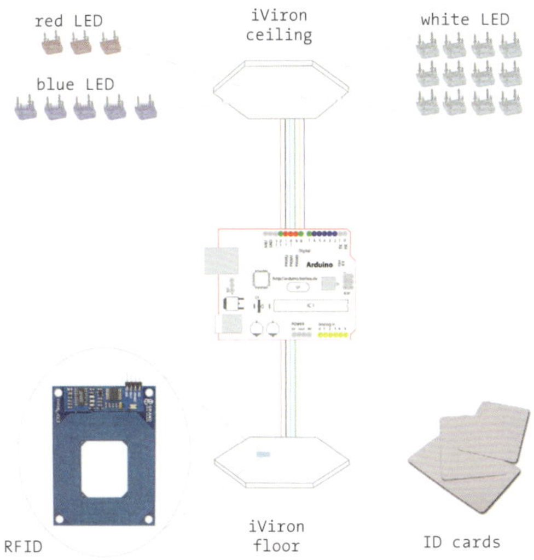

/07

/03
Microclimate interaction and interstitial climates.

/04
The tile strategy includes three types for the various embedded technologies.

/05
Floor prototype. The RFID reader is embedded within the floor tile.

/06
Three scenarios for application. Club, live-work community space and domestic space.

/07
Full circuit including all components.

mediating weathers

Performative third weather spaces are created within the building envelope
/ **Nadya Volicer***

Taking the glass block as a starting point, Mediating Weathers explores the potential of this material to optimise daylight, adjust for solar gain, and convert to a privacy screen by night.

This project is about the 'third weather' present within the home. This is the weather found in the inbetween spaces: between panes of glass, in wall cavities, in root cellars and greenhouses. These spaces are conditioned to have environments distinct from both inside and outside. How can we create third weather spaces which continue to mediate between the exterior and interior, while offering physiological benefits to inhabitants, and spectacle for viewers both inside and out? The glass block has long been one of the most common third weather building materials, housing air in a hollow block of glass which insulates while acting as glazing and structure.

The foliage wall prototype works in the sunlight by using a photocell to trigger a fan, creating a flutter of leaves to filter the sunlight and cast dappled light on the interior. When out of direct sunlight, an internal LED illuminates the block. Night mode tracks the movements of the inhabitant rather than the sun. When a presence is detected by a given block, the fans will be triggered and the leaves flutter. From the exterior, one would observe the movements of the inhabitants as a fluttering of leaves tracking across the surface of the wall.

A second prototype, the fog wall, uses fog to achieve similar outcomes. It relies upon mist within the block to shield from solar gain and provide privacy at night. A third prototype, the rain wall, operates on the principle of the self-sustaining system based on evaporation and condensation. Provided the block is able to ventilate, the nightly condensate would help to cool the living space. By allowing interior night air to pass through this third weather as it moves to the exterior, it would cool the domestic environment. In this model, the built-in privacy screen is the foliage. Two final models harness wind and sun to mediate the third weather space by merging a wind catcher and a double skin façade, opening and closing apertures depending on the season through a grid of concave lenses which supplement the sun's natural path to increase sunlight during the darker days of the year.

/01
Fog wall prototype.

*Guided by Rodolphe el-Khoury and Nashid Nabian

/01

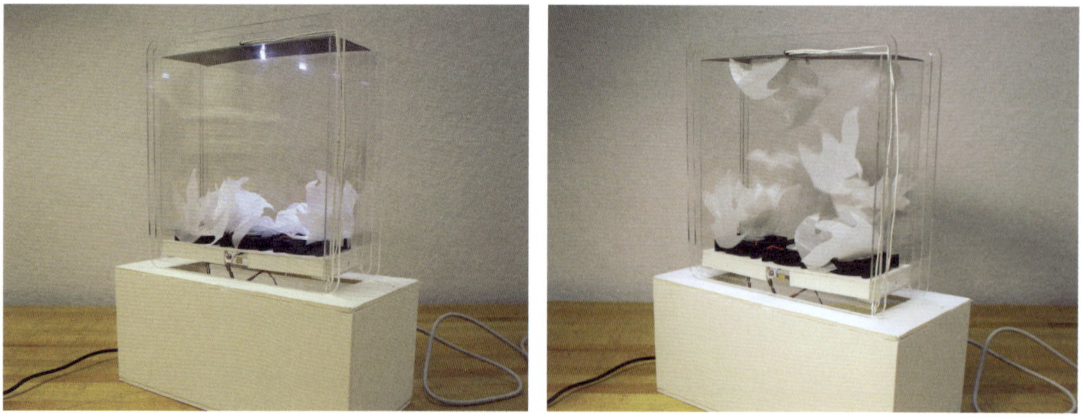

/02

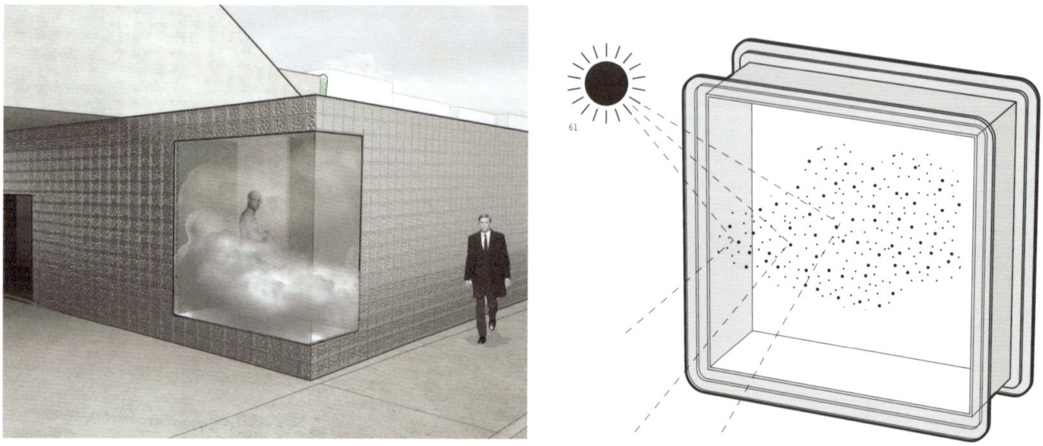

/03

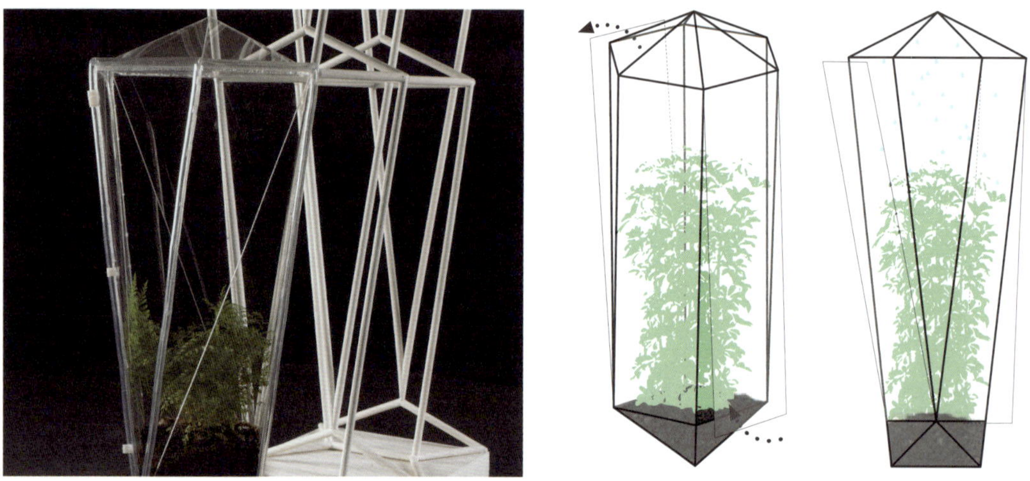

/04

/05

/06

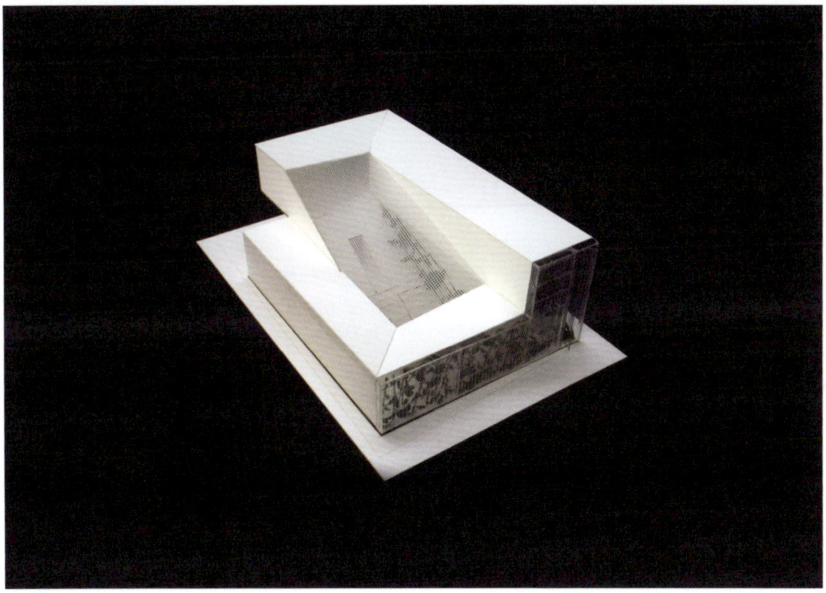

/07

/02
Foliage wall prototype.

/03
Fog wall prototype.

/04
Rain wall prototype.

/05
Sun wall prototype.

/06
Wind wall prototype.

/07
Foliage rain wall in a scale model house.

the sunlight house

A system which uses the structure of the house to transmit light into dark interior spaces / **Yuliya Bentcheva***

The Sunlight House uses a solar tracking device, optical fibres, and light transmitting/emitting material to transform interior spaces using natural light.

With the advance of technology, society is becoming increasingly detached from the natural environment. People spend much more time indoors and in spaces that attempt to recreate the natural environment, and as such, new technologies are attempting to incorporate sunlight into these often dark areas.

The Sunlight House brings sunlight into dark interior spaces through three basic components: a solar tracking device, optical fibres, and a light transmitting/emitting material. The solar tracking device uses a double convex lens in order to focus collected sunlight onto the fibres. These fibres transmit light deep into the house through a light wall, made of a light transmitting or emitting material. This may be glass or plastic and can be incorporated into the structure of the building or used as a light fixture.

/01

/01
Exterior façade built from solar tracking devices.

/02
Section through façade, showing structural details and openings.

/03
Reviving the attic and basement using a light collecting envelope.

*Guided by Rodolphe el-Khoury and Nashid Nabian

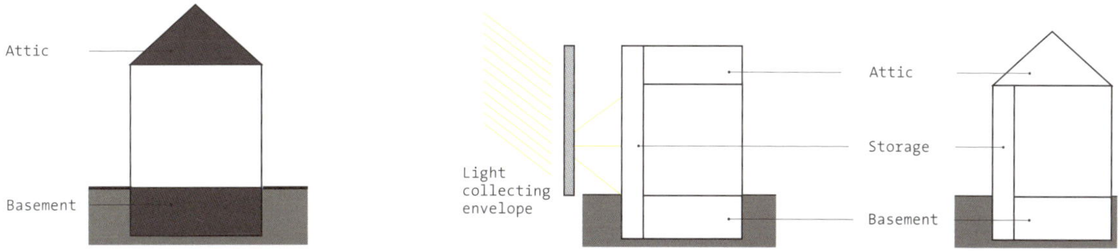
Attic
Basement
Light
collecting
envelope
Attic
Storage
Basement

/03

/02

/04

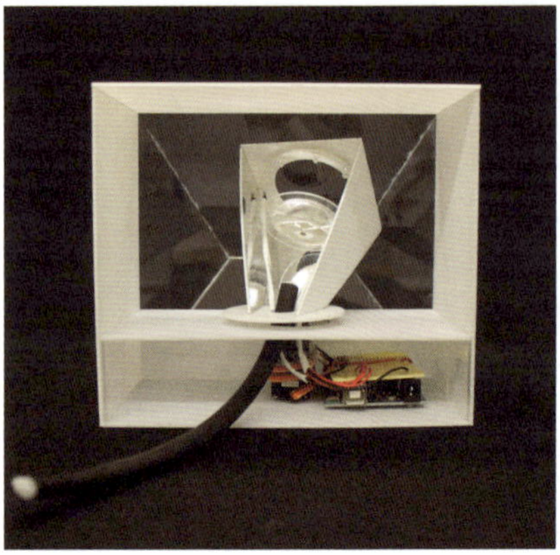

/05

/06

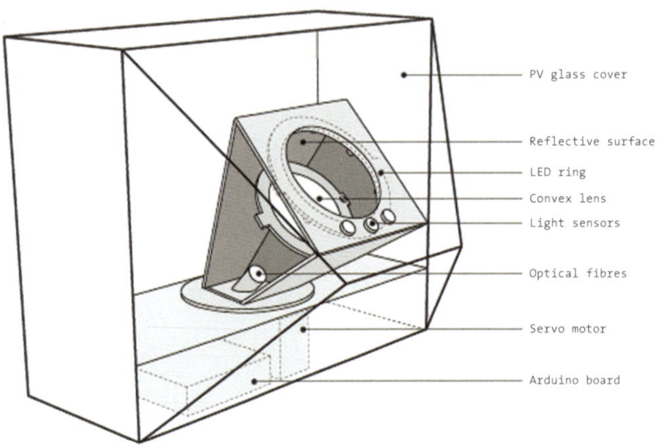

/07

/04
Optical fibre and Arduino setup.

/05
Solar Tracking Device prototype.

/06
Arduino and circuit board.

/07
Solar tracking device assembly.

/08
Light flows from the solar collectors to the cables, and then to the light wall and into the home.

/09
The party wall is expanded to incorporate the core of the house and the circulation. Now that daylight is available throughout the space, the organisation of the home can be reconsidered.

/10
Sectional model of a light wall.

/11
Light wall detail. Light shutters. Light is always behind the shutters and can be adjusted depending on the need in any given space.

/12
Light wall detail. Storage units are embedded into the light wall. When the cabinet doors are opened, light comes out.

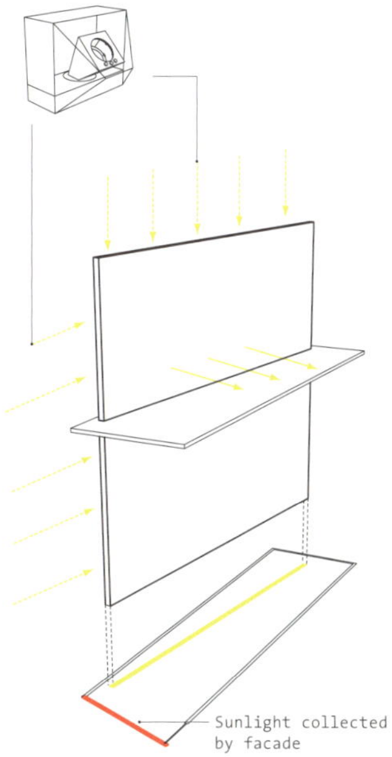
Sunlight collected
by facade

/08

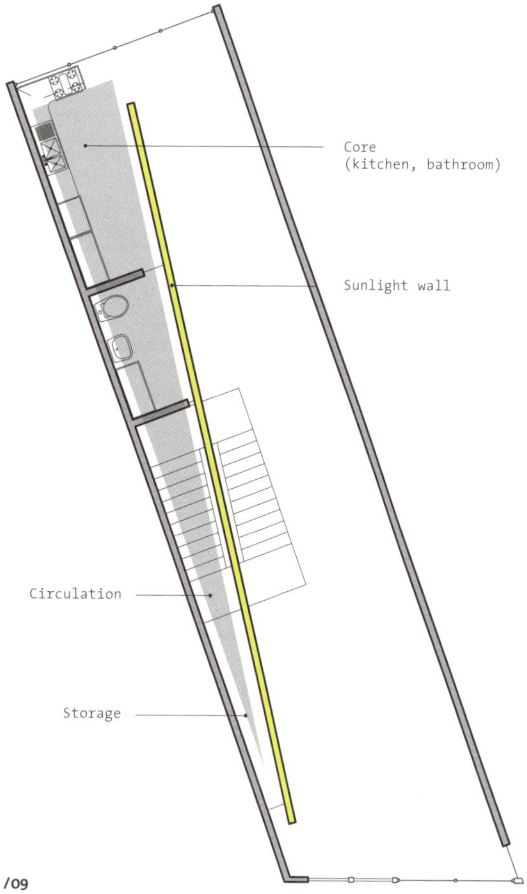
Core
(kitchen, bathroom)
Sunlight wall
Circulation
Storage

/09

/10

/11

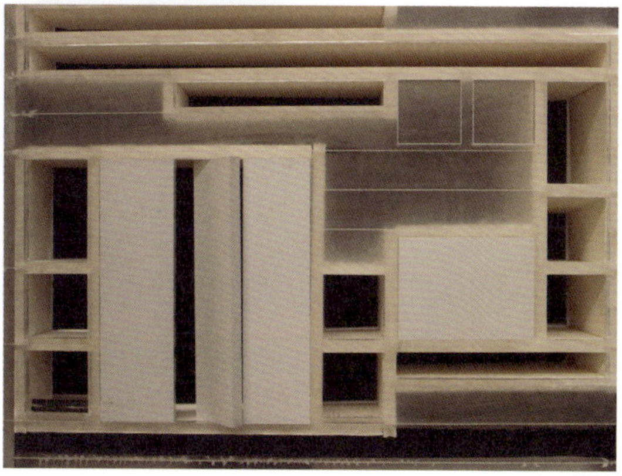

/12

surface as interface

As computational devices merge withthe physical world, displays, keyboards, touch pads, and myriad types of haptic input devices populate the environment. The city becomes an interface that links bits and atoms, its surfaces are embedded with read/write memory. The following projects claim an active role for architecture in linking the flow of information to the physical environment. They build portals or touchdown sites for intensifying info-spheres – a virtual simulacrum of the physical world that increases in magnitude and resolution daily.

IM blanky

IM Blanky is a self-modelling blanket exhibiting primitive cognitive capacities / **Studio NMinusOne with Rodolphe el-Khoury**

A blanket that simultaneously embodies both a physical and digital presence.

The (soft) Hardware

The blanket measures 234.6 cm x 128 cm and is composed of a distributed field of 104 soft tilt sensors. The soft sensor forms the most basic motif: the flower. The flower consists of 6 conductive petals, linked by resistors, and a conductive tassel in the centre. The flowers are grouped together into 14 larger configurations or clusters and 2 half clusters. Working as a directional marker, the tassel's contact with a petal registers a specific orientation or tilt of the blanket. The flowers are arrayed within a circular positive and negative power supply circuit, and their stems plug into a computational hub (Multiplexer). The clusters are then linked together, into a larger network of clusters, each relaying the position of its flowers to a microcontroller stitched to the back of the blanket. (Arduino LilyPad)

The Software

The distribution of sensors is based on an underlying hexagonal structure. Each flower occupies a hexagonal cell, surrounded by six neighbours. As the software receives directional (N, S, E, W) input from a cell, it is able to reconstruct a slope based on the position of that cell and its immediate neighbours, generating essentially a surface of peaks and valleys. (Processing)

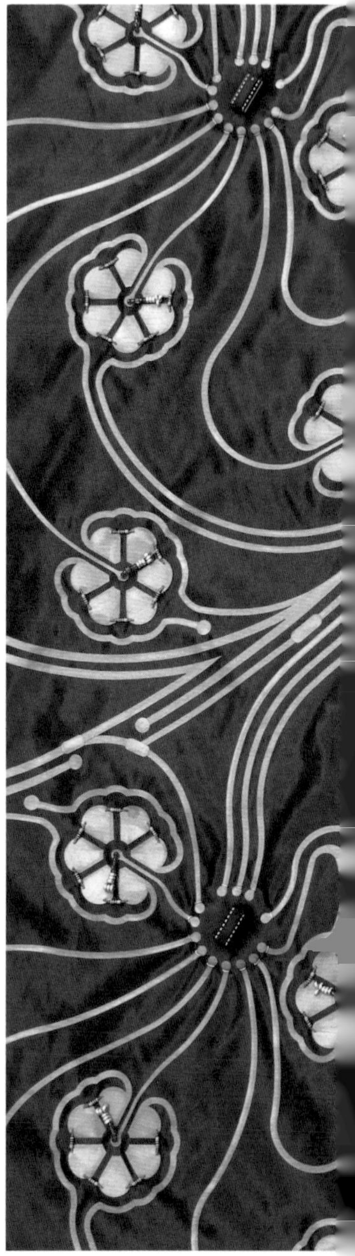

/01

/01
The flowers are grouped together into 14 larger configurations or clusters and 2 half clusters. Working as a directional marker, the tassel's contact with a petal registers a specific orientation or tilt of the blanket. Materials: taffeta, nickel coated silver fabric, conductive thread, silver plated beads, and glass beads as tassel stem insulators.

/02
The first working tilt sensor prototype. Copper fabric, conductive thread and silver beads.

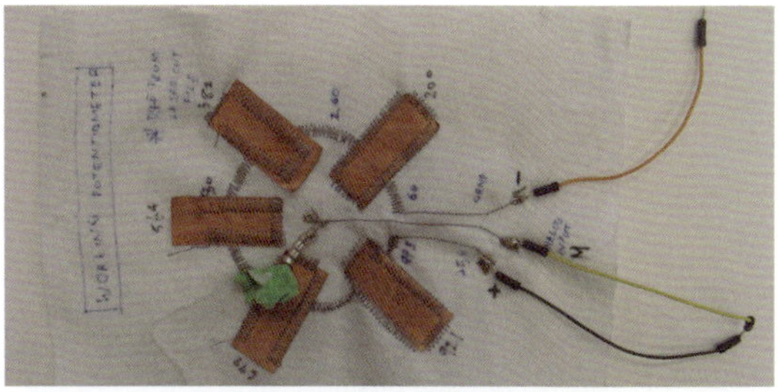

/02

/03
The organic (floral) vs. the orthogonal below. The compact and evenly distributed organisation was ultimately more efficient than the traditional Cartesian circuit layout.

/04
Test patch, stitching circuits with conductive threads.

/05
The evolution of the soft tilt sensor: a search for the most efficient organisation. Materials: conductive fabric and thread, conductive bead.

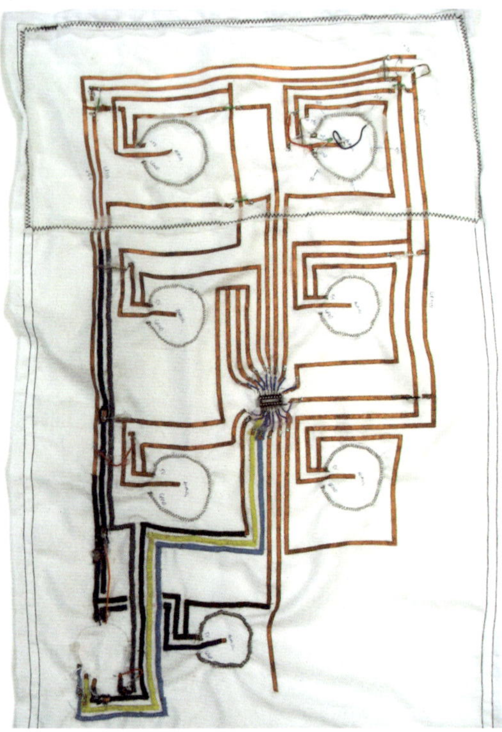

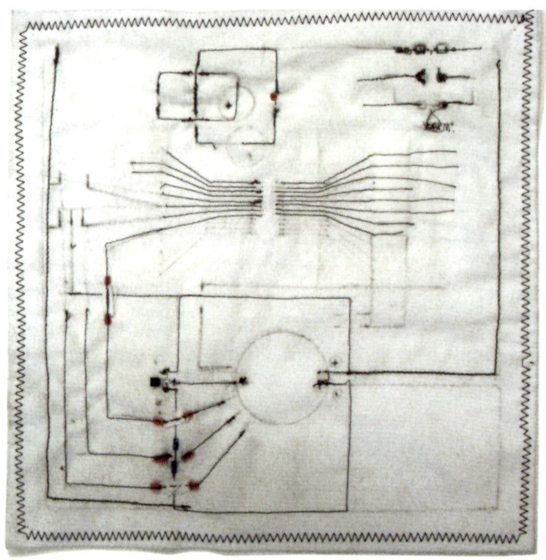

/04

/03

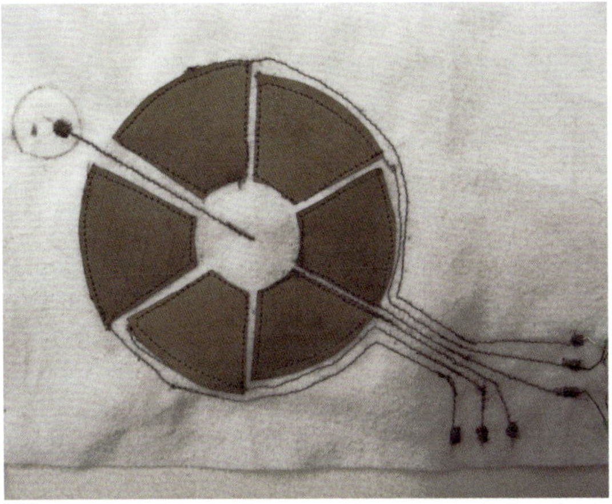

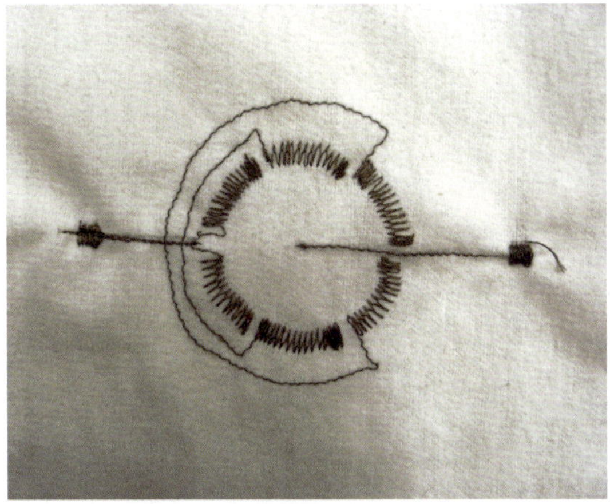

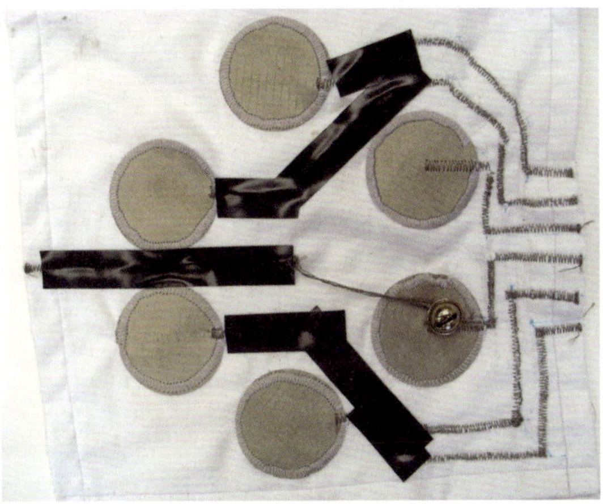

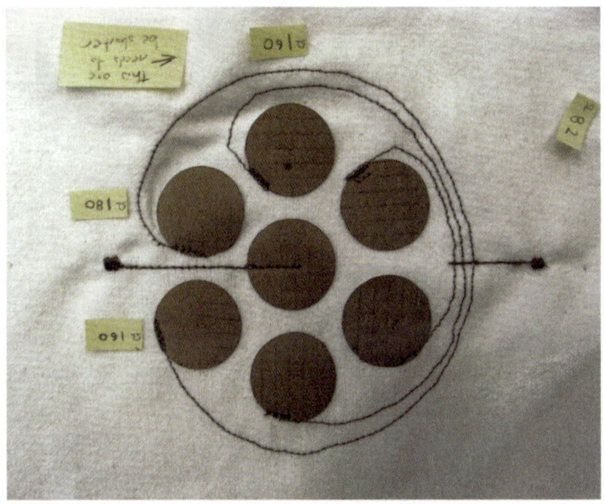

/05

Completed blanket with circuit schematic.

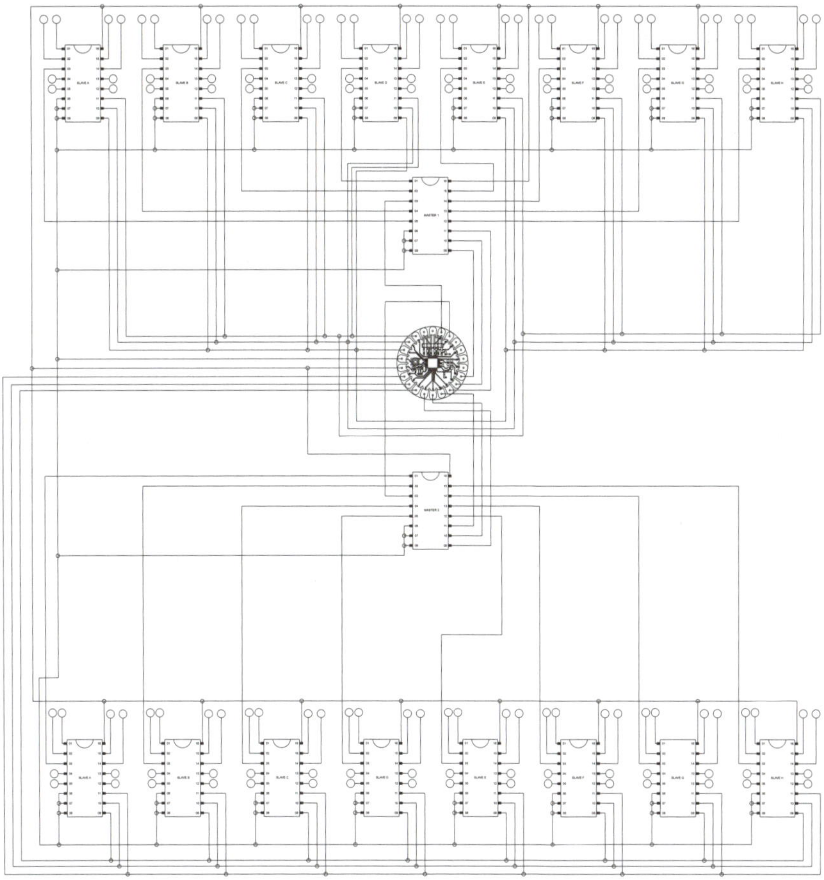

/07

/06
Hubs / Multiplex variations in form and performance.

/07
Circuit diagram.

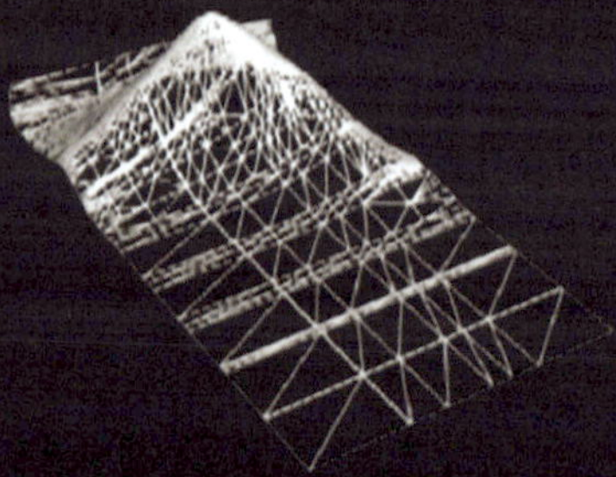

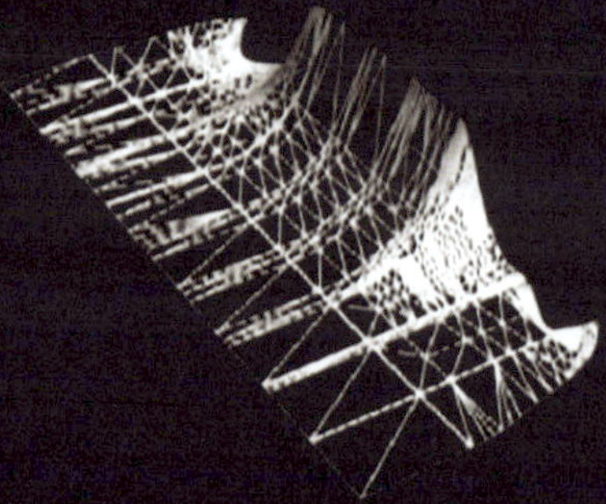

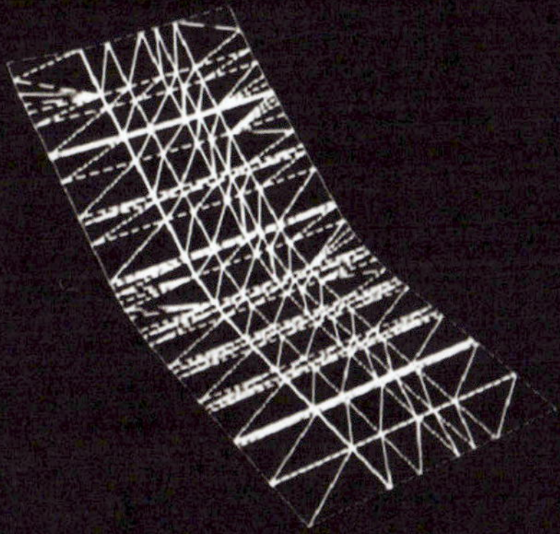

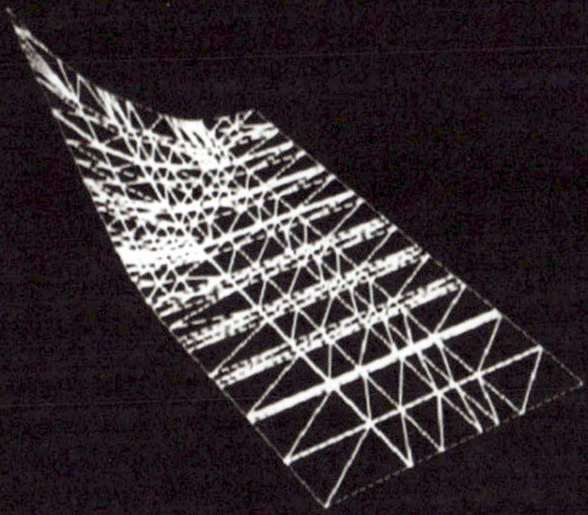

Stills from video depict real time digital replication of actual movement. The blanket simultaneously embodies a physical and digital presence.

light-emitting concrete

Linking the permanence of physical built form with the temporal nature of constantly adapting information delivery / **Ada Juwah***

Light-Emitting Concrete is an electronically enhanced concrete block with embedded dot-matrix LEDs that can be deployed for the erection of substantial structures. In conjunction with user-input recognition mechanisms, Light-Emitting Concrete blocks create interactive, low-resolution information portals at an architectural scale.

"Sometimes I've believed as many as six impossible things before breakfast." —Lewis Caroll, *Through the Looking Glass, and What Alice Found There*

Light-Emitting Concrete provides a linkage between the permanence of the physical built form, and the temporality of real-time information that constantly adapts to new conditions. By combining the dynamic nature of information with the static characteristics of a building material such as concrete, the proposal examines the possibility of creating information architectures consisting of physical spaces of knowledge, exploring the potential of low-resolution display to recombine the analogue and digital aspects of architectural space that is augmented with layers of electronically mediated information.

Light-Emitting Concrete is a building unit composed of glass fibre concrete embedded with dot-matrix LEDs. Walls built with these units provide limited levels of transparency. Meanwhile, they can seamlessly display dynamic information to engage the public. User-input recognition algorithms make this display technology interactive. Interaction is based on the concept of 'me and my shadow', where information is displayed and manipulated within the digital shadow of the user. Cameras capture a live image of the user, which is then displayed on the wall as the user's shadow. This shadow is in turn controlled by the user to select and display information, becoming a window that unmasks hidden information.

*Guided by Rodolphe el-Khoury and Nashid Nabian

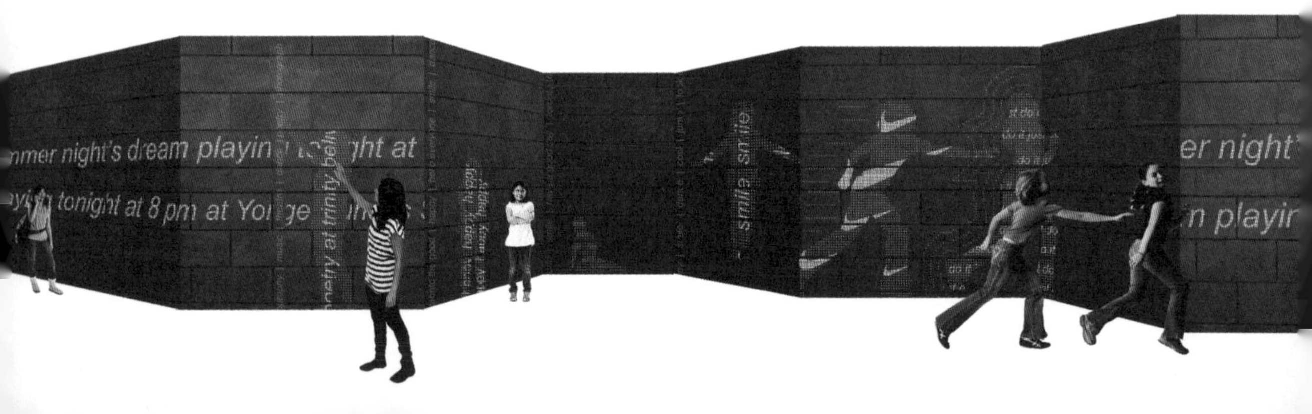

"I wonder if there are any poety readings nearby..."

"Why don't we take a break from studying to see a play?"

ADAPTABLE AND INTERACTIVE INFORMATION DELIVERY

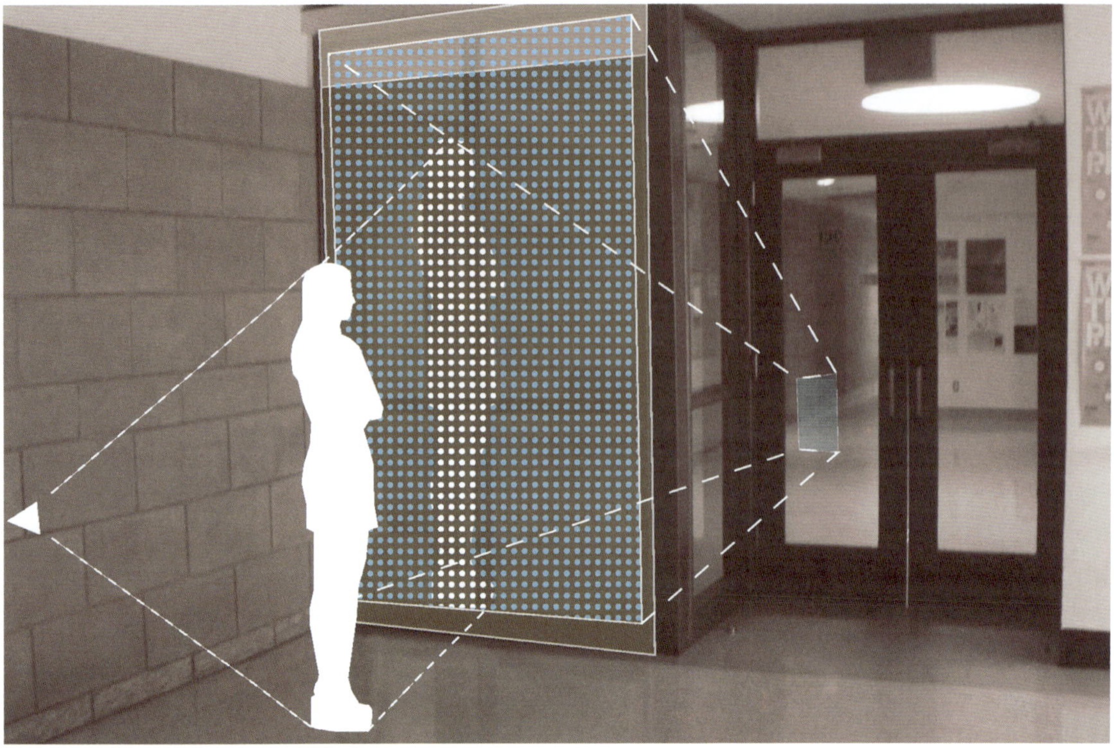

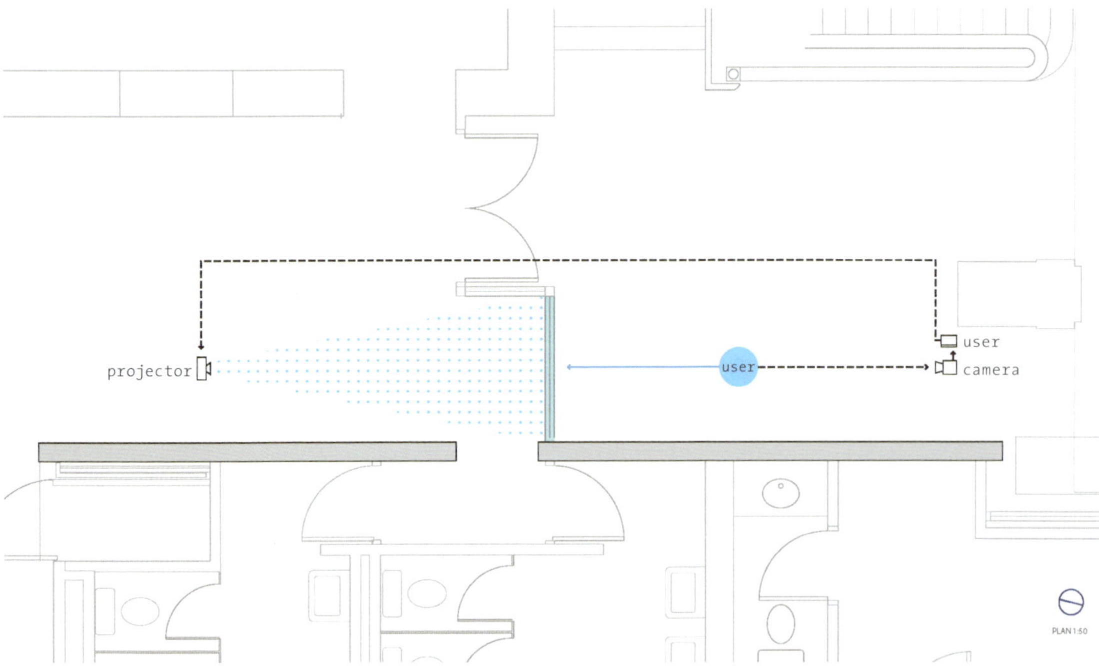

/01

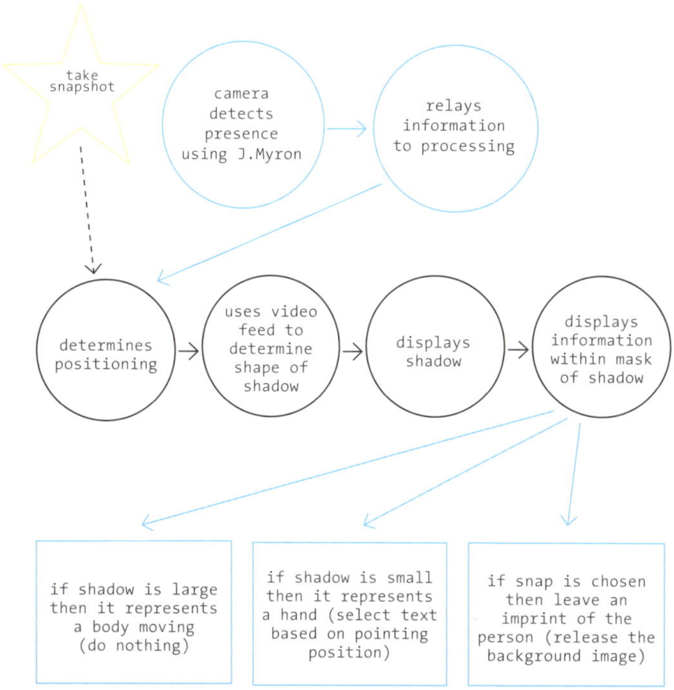

/02

/03

/01
As a proof-of-concept, the wall was created as a temporary display, using a translucent sheet printed with a wall image. The dot-matrix LED information pattern was back-projected onto the wall to complete the effect. A camera captured the movement of the user, which the processing language interpreted to control and display the information pattern in real time.

/02
Processing code.

/03
Users could select which information to display within their shadow, by directing their shadow to touch words scrolling in a ticker to the side.

ambient wallpaper

Bringing attention to the noise level of the room and encouraging users to decrease their sound level / **Azadeh Mohammadi and Taraneh Meshkani***

Ambient Wallpaper is an interactive ornament. The 'transactive intelligence' that enhances the performance of this ornamental element allows it to respond to human activity and influence human behaviour in return. By changing visual characteristics, it provides visual hints as to the proper code of conduct within an architectural space.

Ambient Wallpaper examines the potential of technology to develop dynamic spaces which have the ability to interact with users and communicate implicit messages to them. Ambient Wallpaper is digitally enhanced with 'transactive intelligence' to respond to changing conditions by modifying its visual characteristics, including its colour and pattern.

For prototypical implementation, an interactive applet running in the processing environment on a computer is projected onto the architectural surfaces. The colour of the projected wallpaper changes with the sounds of the environment. With this effect, the wallpaper brings attention to the noise level of a room and encourages users to decrease the volume of their conversations. Aside from detecting noise levels with the use of sound sensors, the wallpaper also responds to the movement of users in the space by deploying multiple distance sensors installed on the projected wall. An Arduino board collects the information from distance and sound sensors and communicates it to the processing environment. Patterns of Persian tiles are used as a model for the design of the wallpaper scheme. The geometrical nature of the chosen Persian pattern allows for the parametric modification of its visual effects by using computational algorithms triggered by detected changes within the context.

/01

/01-02
Demonstration and diagrams showing wallpapers' pattern responding to presence and movement. The Islamic-ornament inspired pattern unravels into an organised chaos of increasing irregularity.

*Guided by Rodolphe el-Khoury and Nashid Nabian

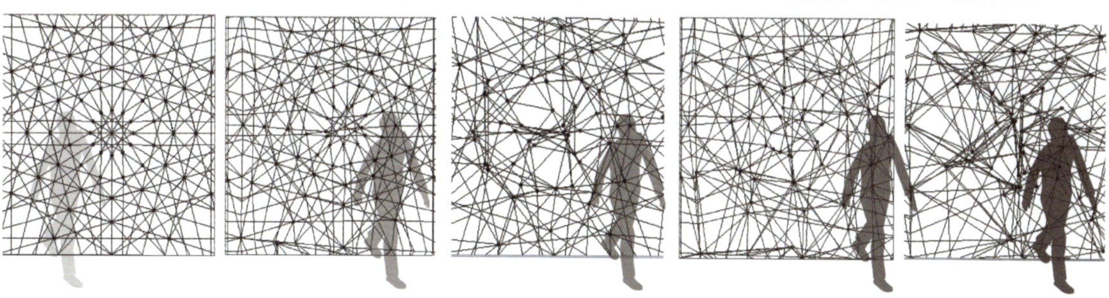

/02

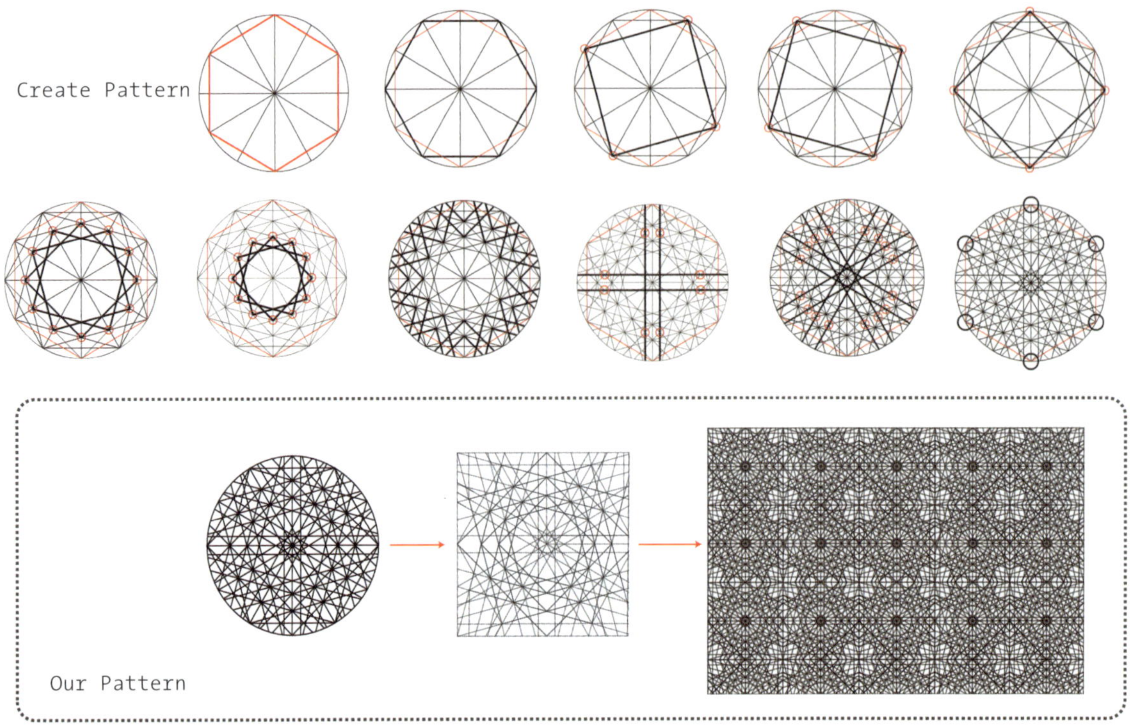

/03

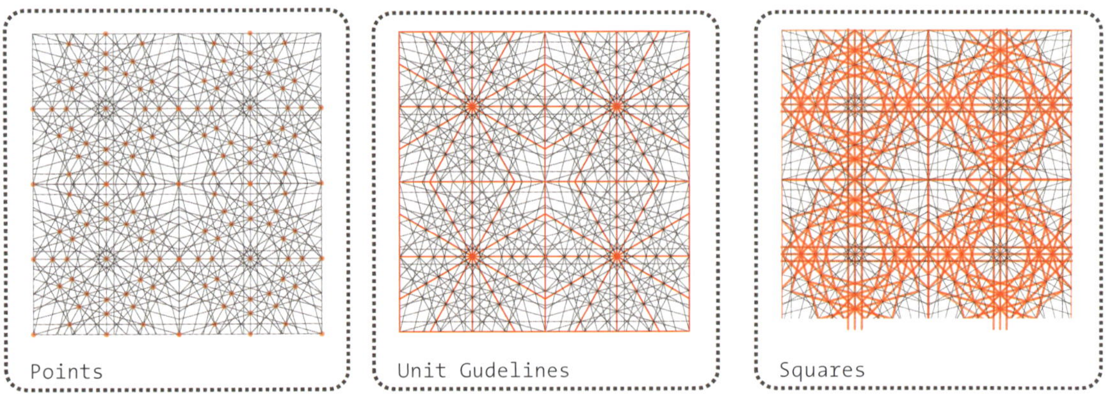

/04

/03
Genealogy of pattern geometry rooted in Islamic traditions.

/04
Highlighted control points within the responsive pattern.

/05-06
Diagrams showing installation and configuration of micro-controlled sensors and actuators.

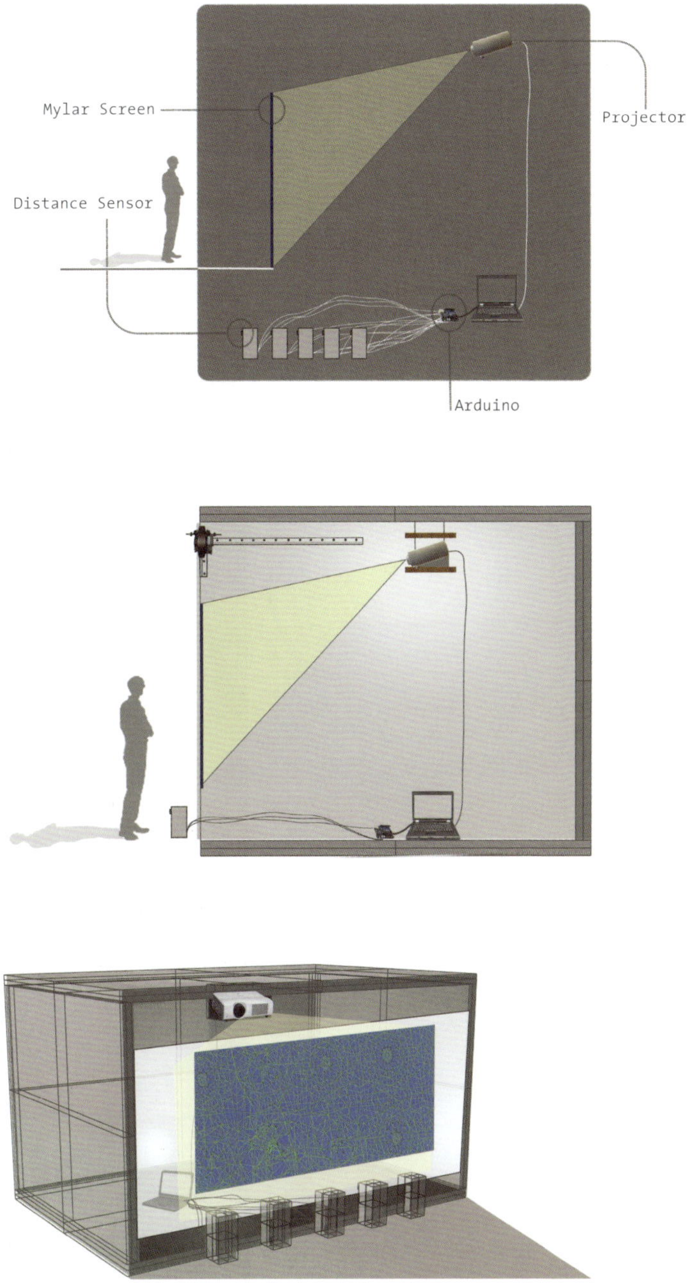

/05

/06

digital window

A window which digitally displays the ‘other side’ / **Studio NMinusOne**

The Digital Window produces a digitally mediated transparency. The Digital Window appropriates a common hardware application found in popular image capturing devices.

The effect accidentally produced by these devices is a digitally mediated transparency, or invisibility. It is this ‘side effect’ that the digital window opportunistically assimilates. The digital window is a window which digitally displays the ‘other side’. Besides relaying the ‘other side’, the window can display a variety of other things: TV, the Internet, or the ultimate reality TV. Similar to the one-way mirror, one can only look out from the window, but not in.

/01
Window as camera and screen.

/01

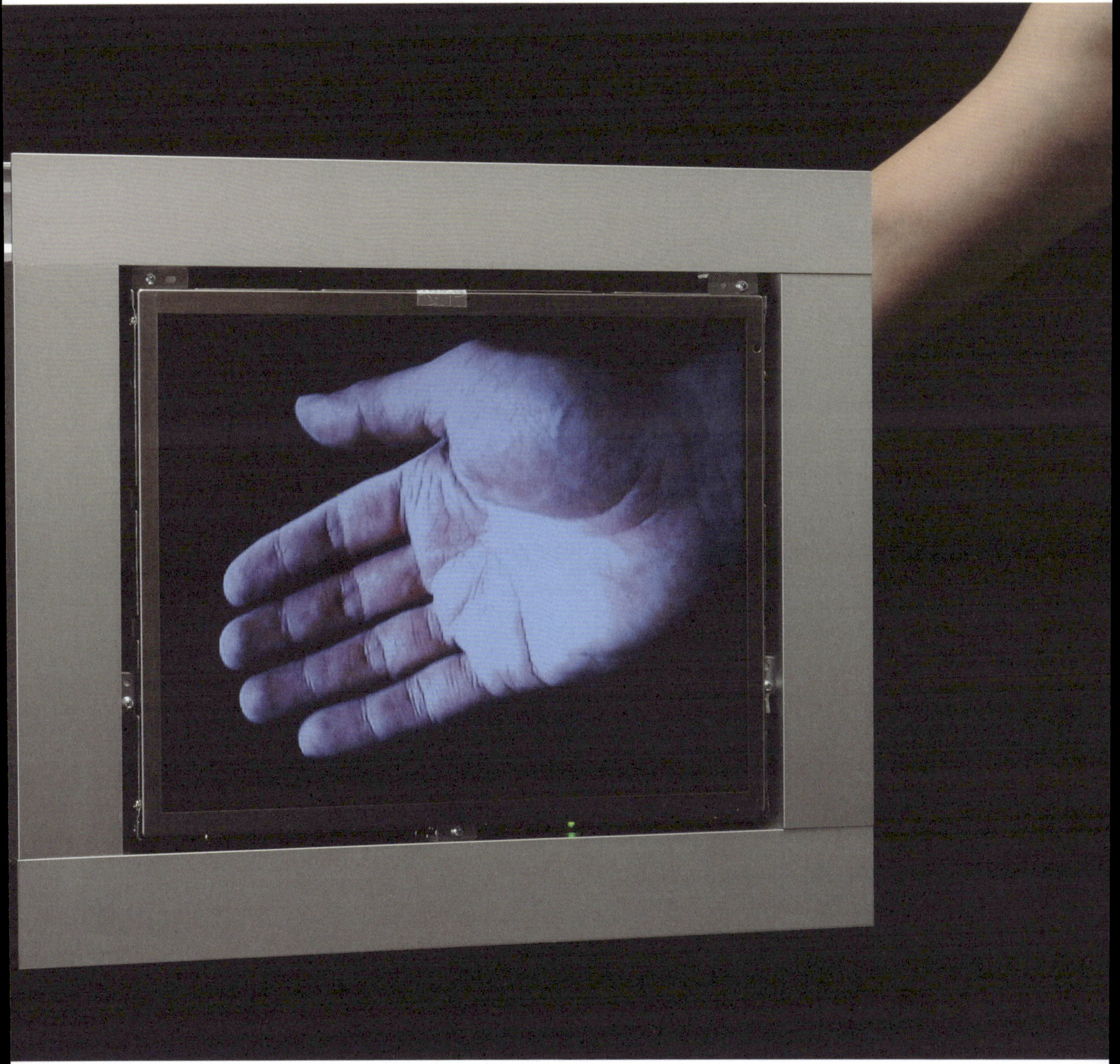

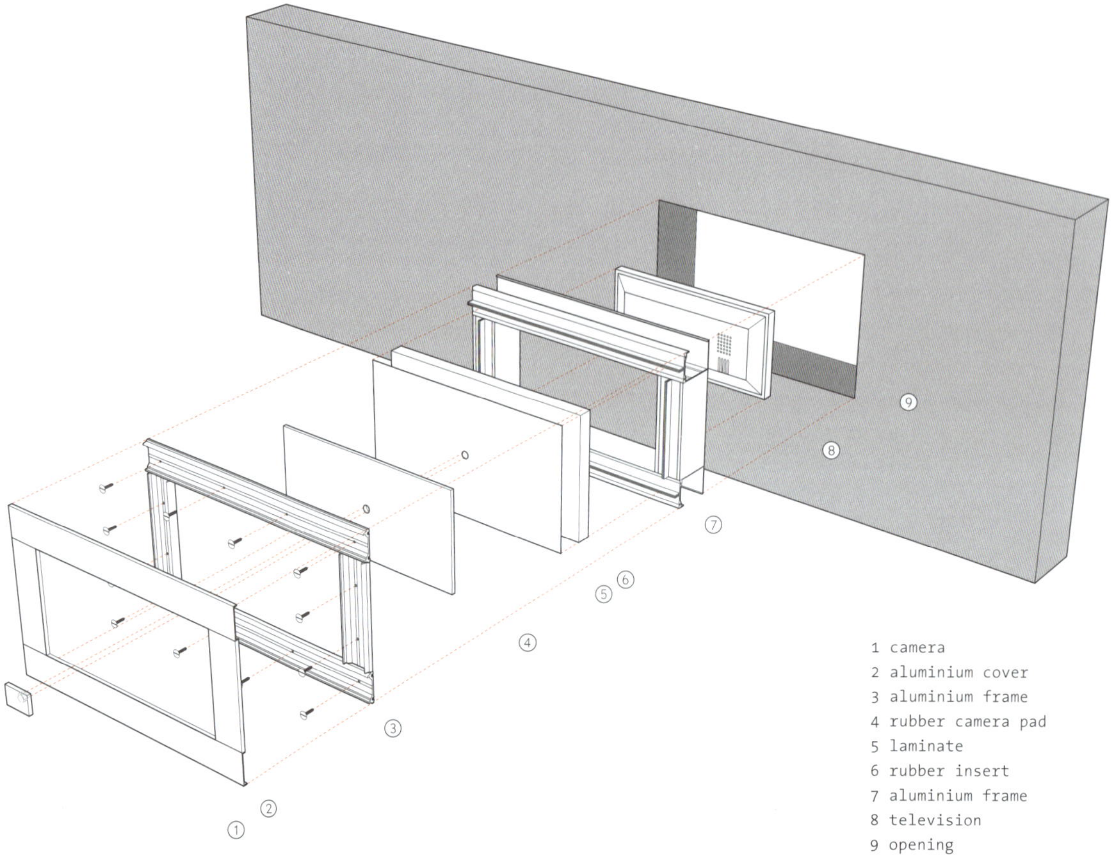
1 camera
2 aluminium cover
3 aluminium frame
4 rubber camera pad
5 laminate
6 rubber insert
7 aluminium frame
8 television
9 opening

/02

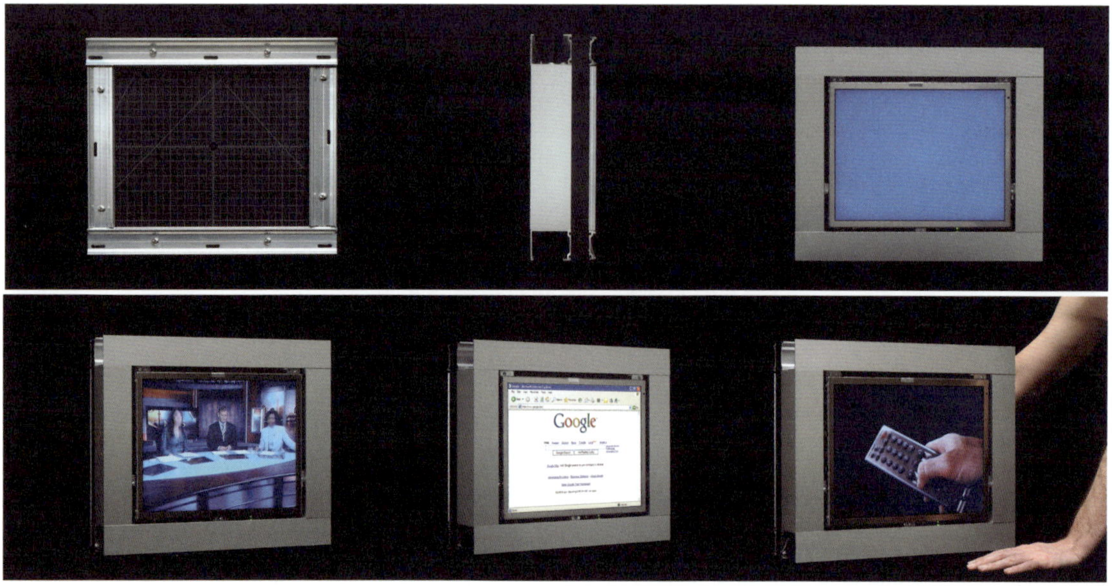
Google

/03

/04

/02
Digital window assembly.

/03
Exterior, side and interior surface.

/04
Different display scenarios.

window 2.0

A responsive system dynamically adjusts the perspective of the virtual space according to the position of the viewer / **Khoury Levit Fong**

Window 2.0 uses responsive video imaging to digitally augment physical space by tracking the position of the viewer and adjusting the virtual space accordingly.

Window 2.0 is designed to simulate, enhance or entirely reconfigure the optics and the visual phenomenon of a physical window. Applications range from teleconferencing and telepresence to virtual spatial extension and transparency.

Window 2.0 produces responsive perspective effects by means of two methods:

Window 2.01. The position of the viewer is pinpointed by means of video tracking with a webcam. It dynamically controls the orientation of the viewing camera mounted on a servo motor. The responsiveness of the camera can be finely calibrated to react very sensitively to minute signals in body language, shifting the view in response to a slight shoulder rotation or to more closely imitate the interaction of viewers with physical windows in real space.

Window 2.02. The position of the viewer, tracked with a webcam, controls the virtual perspectival orientation of the live video feed by means of digital processing.

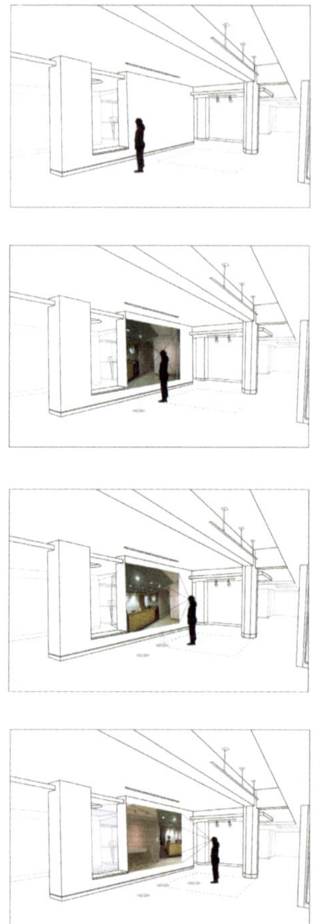

/01

/01
Diagram of responsive behaviour.

/02
Still from video demonstration.

/02

❶ The coordinates are translated into a relative angle which is passed from the CPU to Arduino to a servo on the other side of the wall

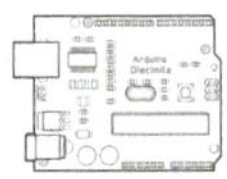

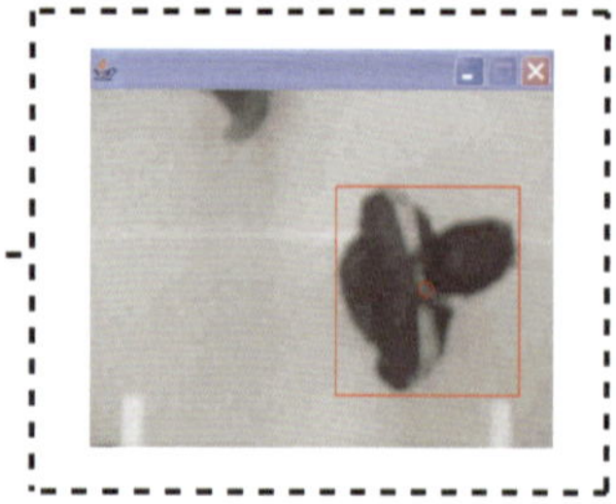

❷ The servo rotates a mounted camera to create a virtual line of vision betweem the user and the space behind the wall. This camera's video feed is sent back to the CPU

❸ The video feed is then analysed using Processing to determine the x and y coordinates of the user relative to the projection screen

/03

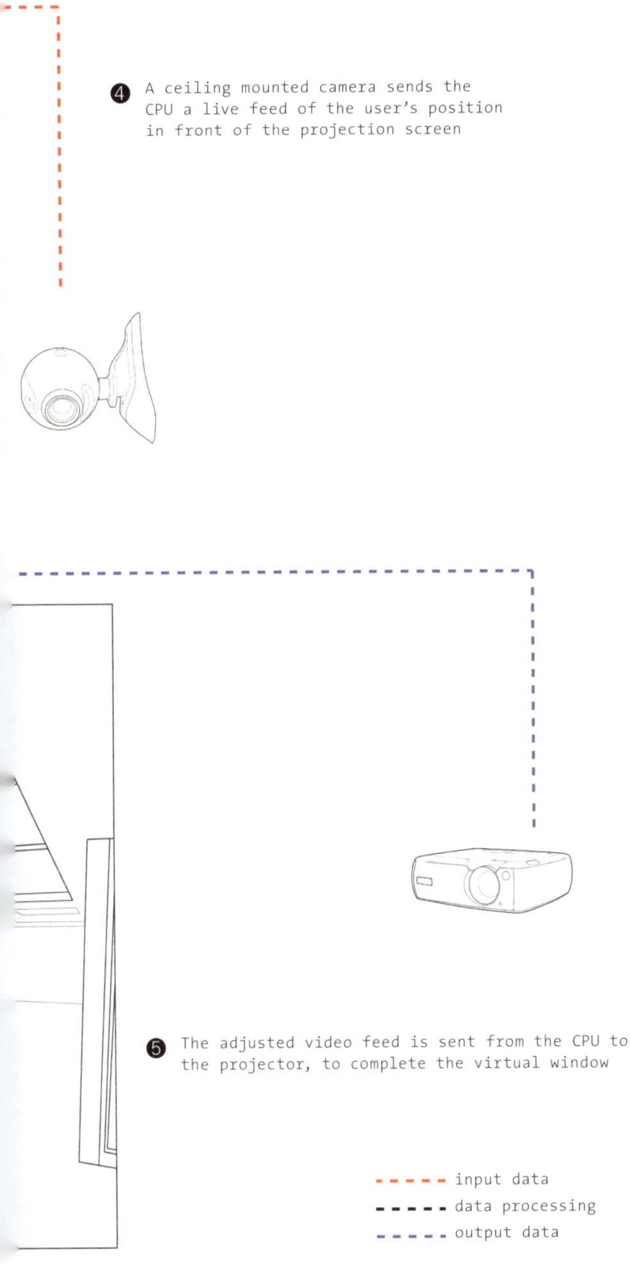

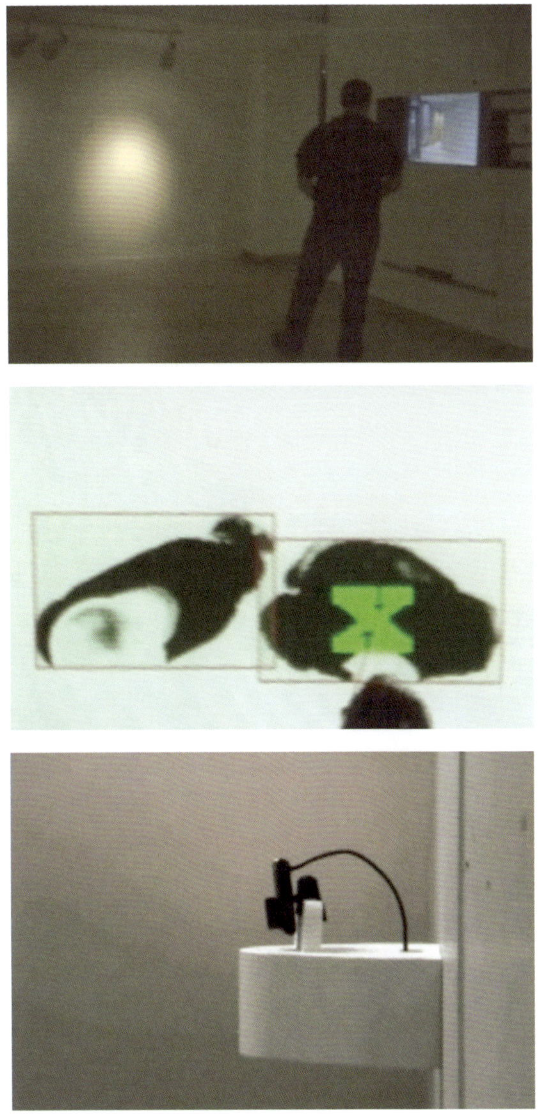

/04

/03
Diagram of video tracking, and camera/servo control.

/04
Video stills capturing the three main elements of the device in operation: the interaction of the user with the window/ screen, the video tracking and locking process and the servo-mounted webcam.

MyCity

MyCity aligns architecture with information technology to empower the sensate body and its spatio-temporal envelope as the intimate locus of social connection / **Khoury Levit Fong with Mark Jones (Emergent Software)**

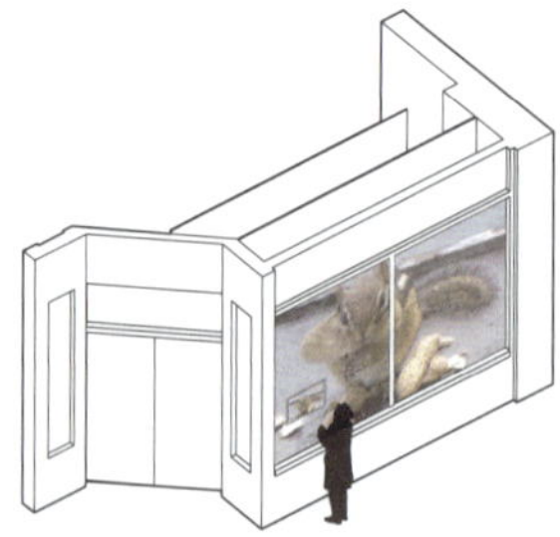

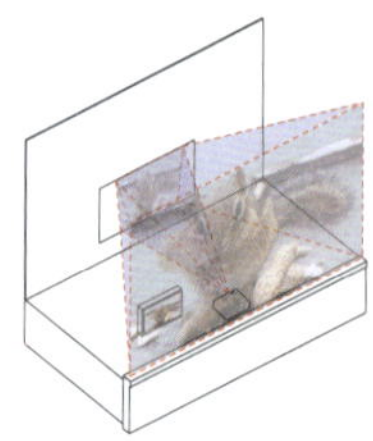

MyCity explores the capacity for environments to respond pointedly to a subject's actions and characteristics.

The individuated connectedness is meant to expand the subject's sphere of influence, projecting in this case, the personal Flickr gallery of any willing participant onto public space. This is achieved simply with the help of a cell phone that allows an individual to identify his or her Flickr account and trigger the public display of selected personal images.

MyCity encourages individuals to translate their web-presence into a physical reality allowing them to appropriate public space for their own use and expression.

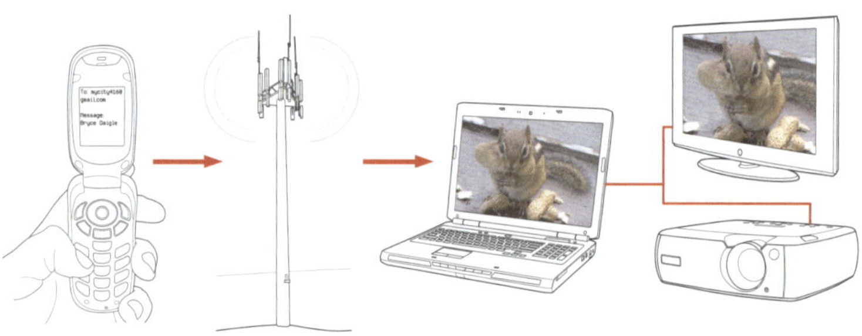

/01

/02

/01
Diagram explaining the mode of user participation.

/02
View of installation with rear projection in operation.

NO
STANDING
mycity416@gmail.com
flickr

Aleph TO

A projected image mirrors the scene on the street / **Khoury Levit Fong with Nashid Nabian**

Aleph TO uses digital projection to create a responsive image overlooking the street. Triggered by the proximity of the viewer, it is constantly changing to reflect the scene on the street.

Aleph TO's resolution varies with pixels diminishing or augmenting in size depending on the proximity of the viewer. Each pixel is a photograph that is grabbed in real time from Flickr galleries on the web. The continuously updated photographs, identified and selected by keywords in their tags, capture concurrent scenes throughout the city.

"'When I opened my eyes, I saw the Aleph.' 'The Aleph?' I repeated. 'Yes, the only place on earth where all places are seen from every angle, each standing clear, without any confusion or blending.'"—Jorge Luis Borges, *The Aleph.*

/01

/01
The resolution of the image changes in proportion to the distance of the viewer from the mirror/screen measured by motion and proximity sensors.

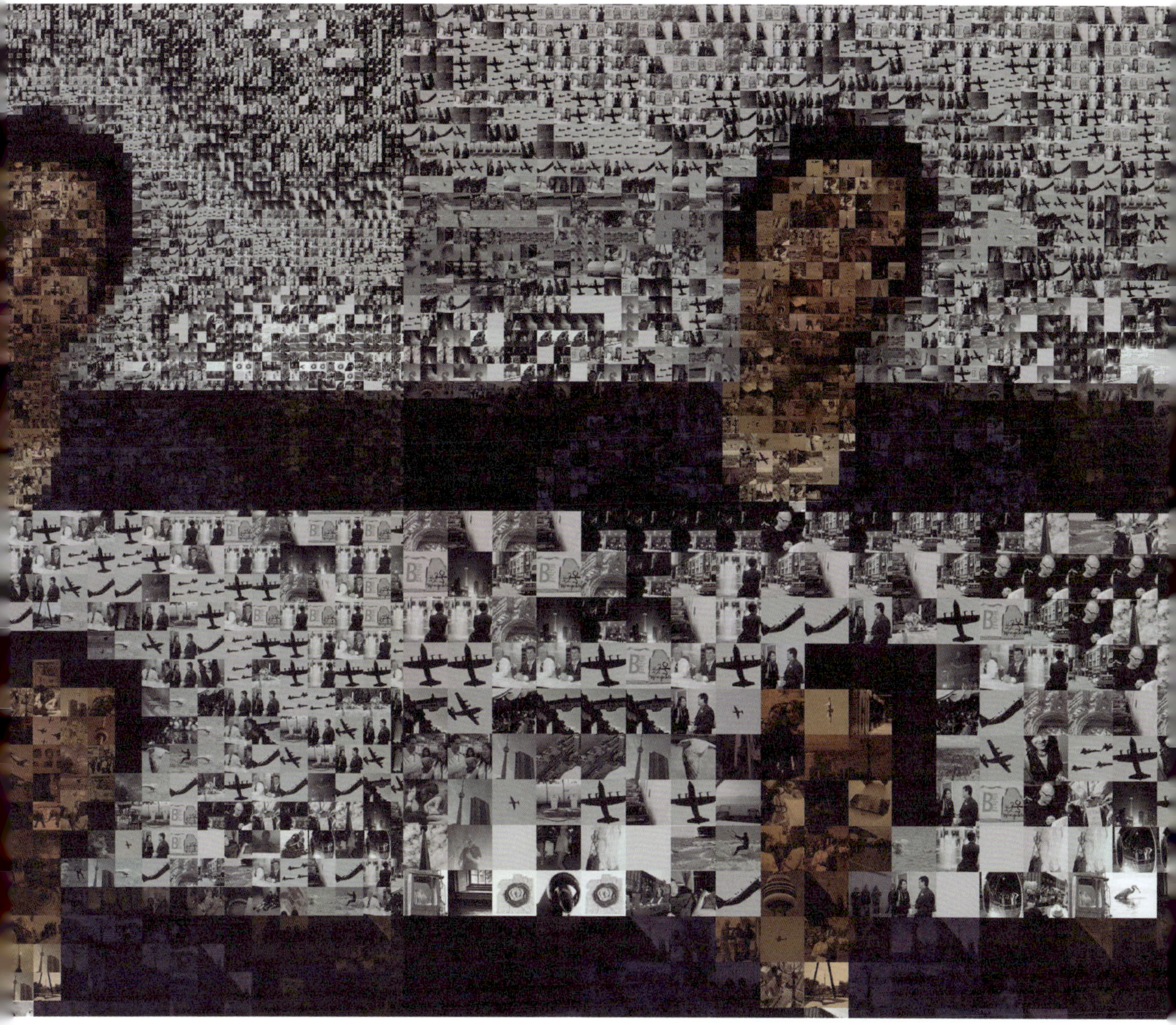

Aleph-TO

View of the installation in operation during Toronto's 2009 Nuit Blanche event.

linking shade

Strung ceramic beads offer shading and dynamic lighting citywide / **Khoury Levit Fong with Kuth Ranieri Architects**

Football-shaped tiles, designed to be strung like beads on string are suspended over public space in a variety of structural configurations and shading scenarios in Phoenix, Arizona.

The strung ball-tiles can hang by cables from concrete columns along sidewalks as well as in park settings. They can also shade public alleys or small urban squares by attaching directly to the sides of buildings. The perforated ball-tiles come in a variety of designs and colours to achieve a range of compositional and environmental ambitions. Variations include embedded photovoltaic cells, motion sensors, and LEDs allowing tiles to function as autonomous and responsive light sources. At night, the tiles glow in patterns of colour controlled interactively by sensors responsive to movement and any number of programmable inventions.

The materiality of the ceramic ball – which includes recycled ingredients – and their simple moulded production coincides with the economy and simplicity of vernacular traditions of moulded roof tiles. It is able to draw upon experiences of common familiarity while doing so in a manner that refashions a familiar material according to up-to-date and emerging technologies – technologies which are themselves reshaping our experience as an urban public.

/01

/01
Three deployment scenarios for varying urban conditions.

/02
Prototype of ceramic tile.

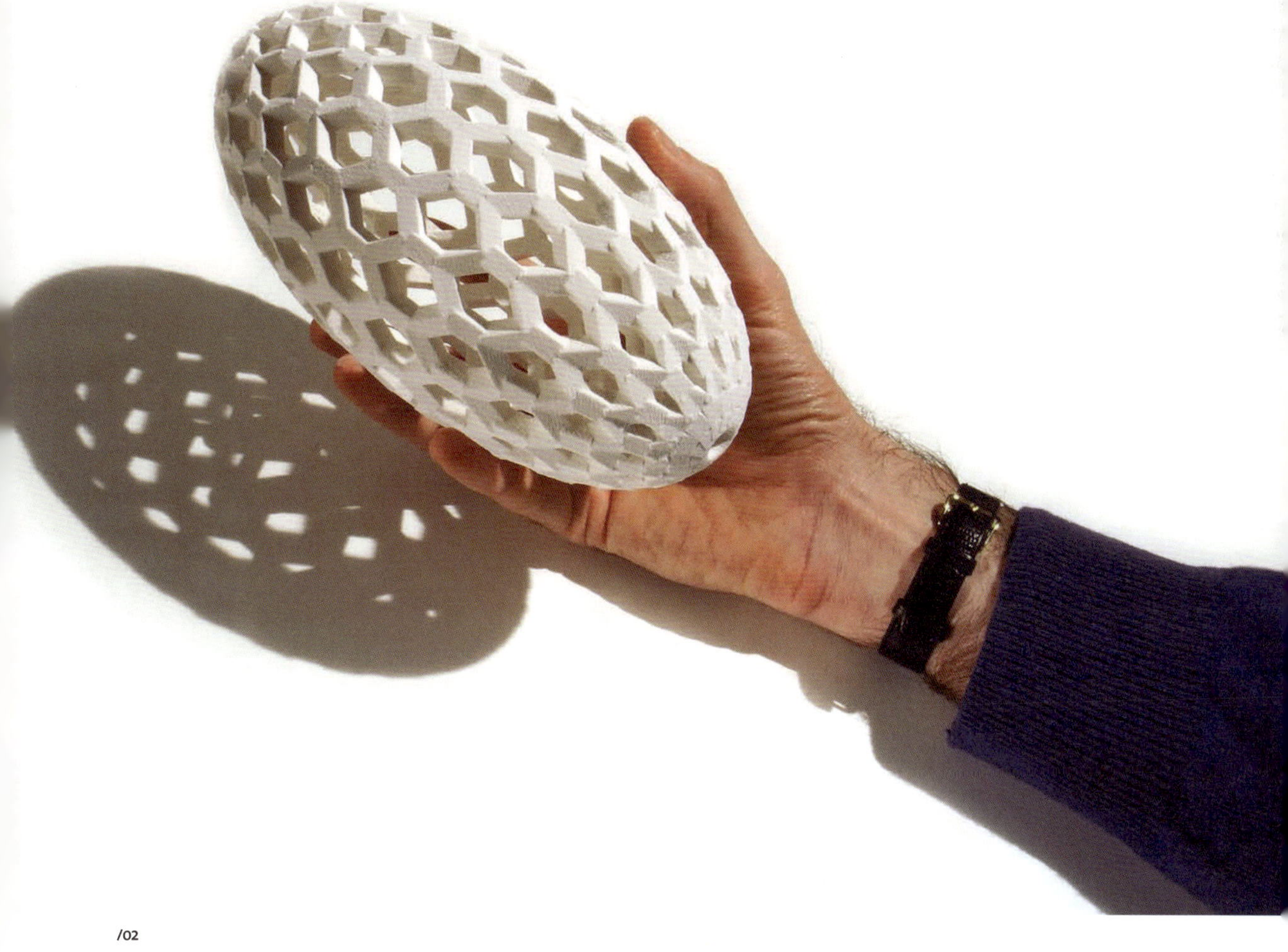

/02

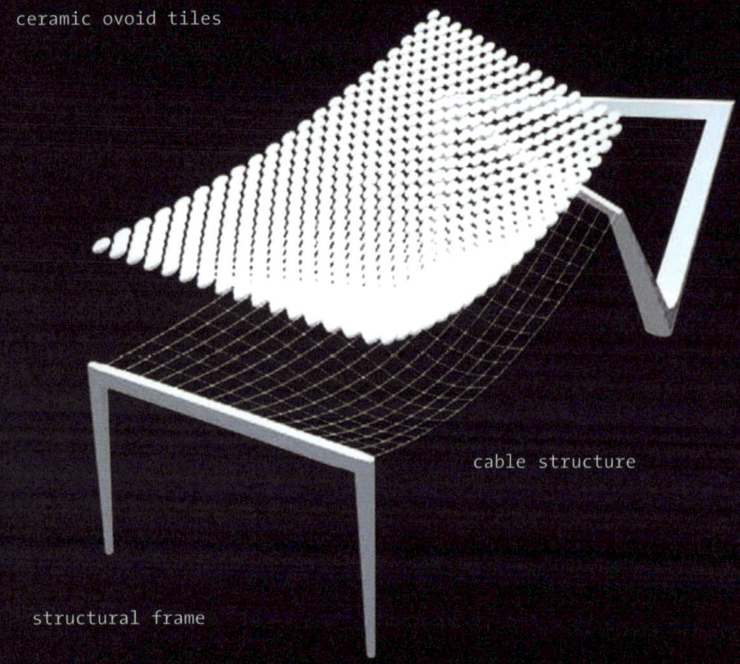

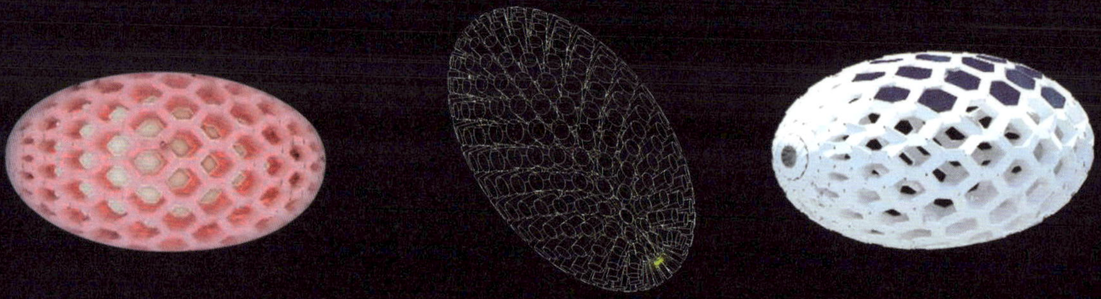

/03

/04

/03
Structural frame and tiles showing embedded LEDs and PV cells.

/04
View of public space under the canopy.
The material character of the ceramic ovoid tiles and their molded production evoke the economy and simplicity of vernacular molded roof tiles thus refashioning the familiar according to up-to-date and emerging technologies.

/05
Prototype of ceramic tile.

/05

Stratford square

A suspended programmable net forms a virtual roof for a public square
/ Khoury Levit Fong

The proposed canopy for Market Square in Stratford, Connecticut is a prominent architectural element that shelters a bus-stop, acts as a digital roof/screen, and mediates sound.

The canopy, made of wood, is integral to a tensegrity structure that is deployed above the square to form a virtual roof: the primary purpose of the steel-cable net is to suspend the wood canopy while the canopy's compressive members, conversely, serve to anchor and stabilise the steel net.

The steel net holding the canopy also supports an array of fibre-optic cables. It is designed, along with auxiliary fixtures, to light the square in a variety of programmable patterns. Possible lighting scenarios include interactive web-based programming and sensor triggered real-time responsiveness.

By means of embedded transducers, the wood canopy performs like an oversized loudspeaker. It amplifies the sound of the cascading water in proportional response to the ambient city noise.

/02

/01

/01
View towards Market Square.

/02
Various programmatic activities are accommodated under and around the net, enhancing public use.

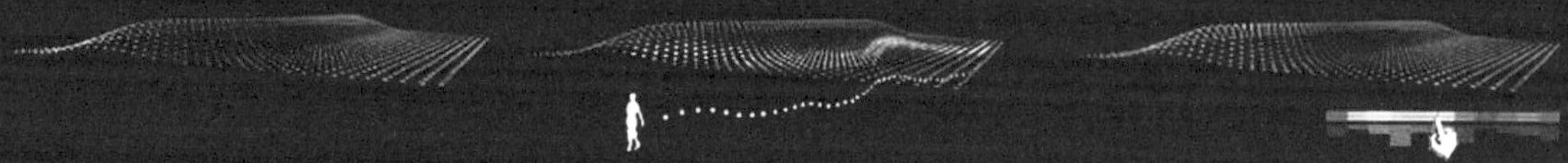

/03
The colour adapts and gradually shifts for a variety of effects and purposes such as weather forecasting or audiovisual spectacle.

/04
The light intensely responds to pedestrian traffic by means of motion sensors.

/05
The colour scheme can be decided through an interactive website.

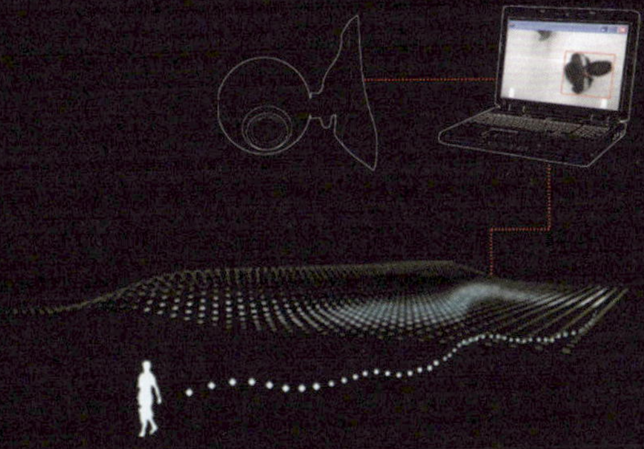

/03

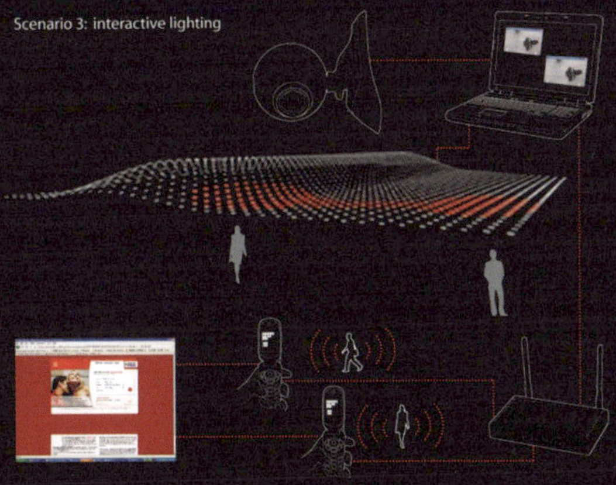
Scenario 3: interactive lighting

/04

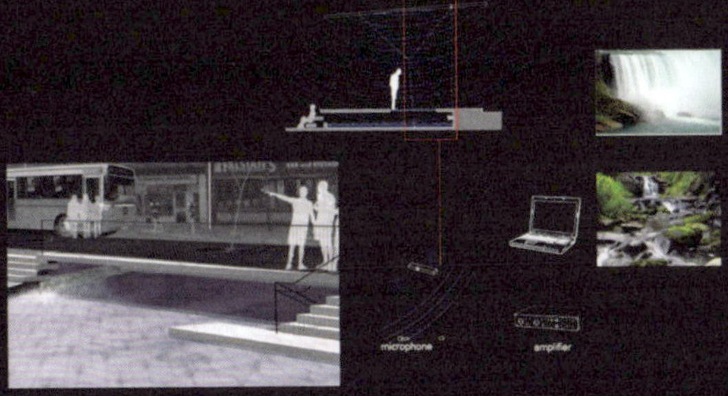
microphone
amplifier

/05

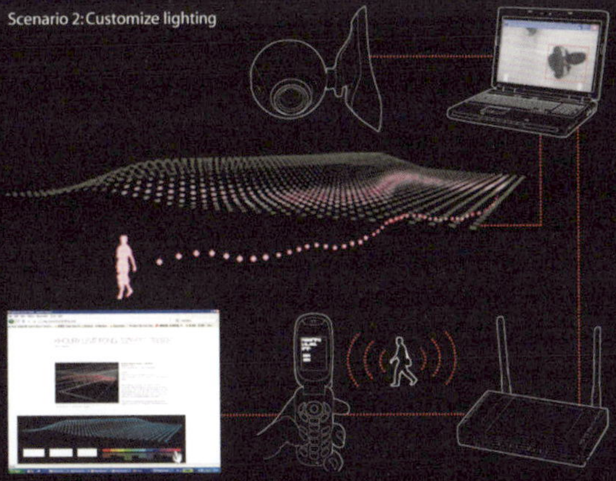
Scenario 2:Customize lighting

superblock

Embedded LEDs and wireless communication transform structural glass blocks into a variable resolution display / **Khoury Levit Fong with Nashid Nabian**

A standard 8 x 8 LED array is embedded into a generic glass block to form a digital display. The image displayed is received wirelessly by means of an embedded ZIGBEE device.

The LED pixels are extended with an egg crate-like diffuser to cover the entire face of the brick. The Superblock can function alone as a low resolution 64-pixel display or can be aggregated with other blocks in a variety of configurations and scales. When Superblocks wirelessly detect and situate other adjacent blocks they re-map the image over the entire field that is constituted by the aggregated blocks. The design and wireless connectivity are meant to enable adaptation to different conditions and facilitate installation. This is a plug-and-play product that can be used for building wall partitions at different scales and in a variety of spatial and structural applications.

/01

/01
A wall of Superblocks in a domestic interior.

/02
Image resolution and dimensions wirelessly and automatically adapt to block configuration.

/02

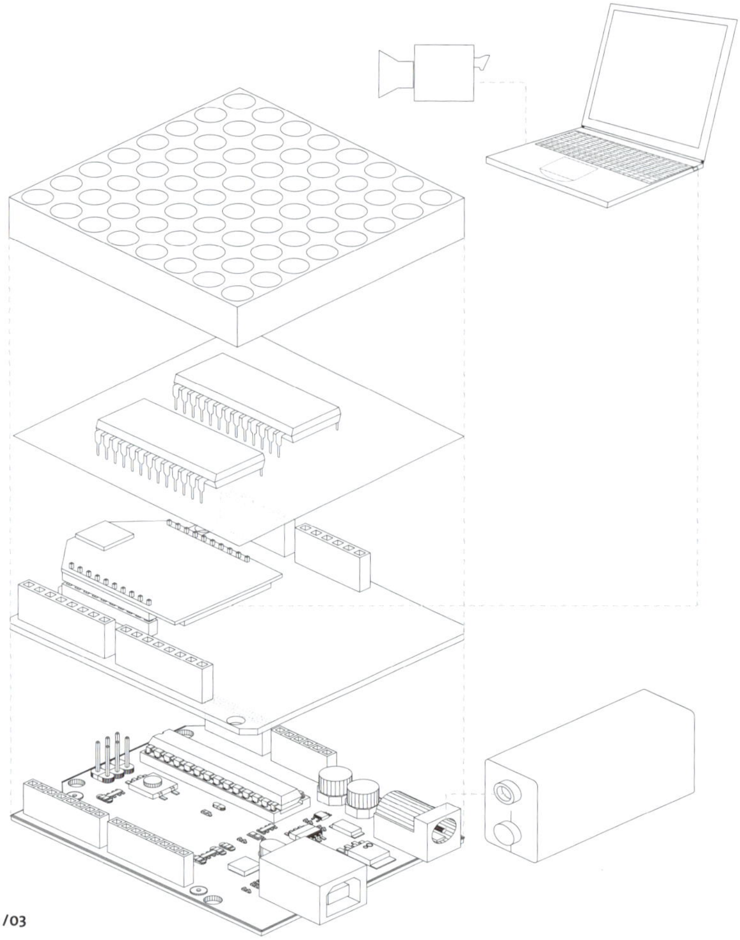

/03

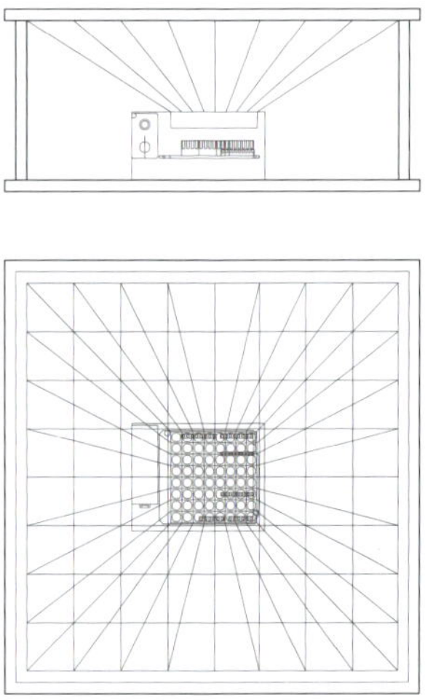

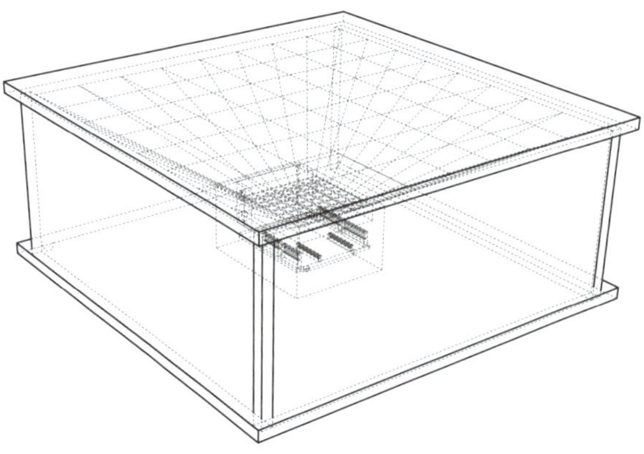

/04

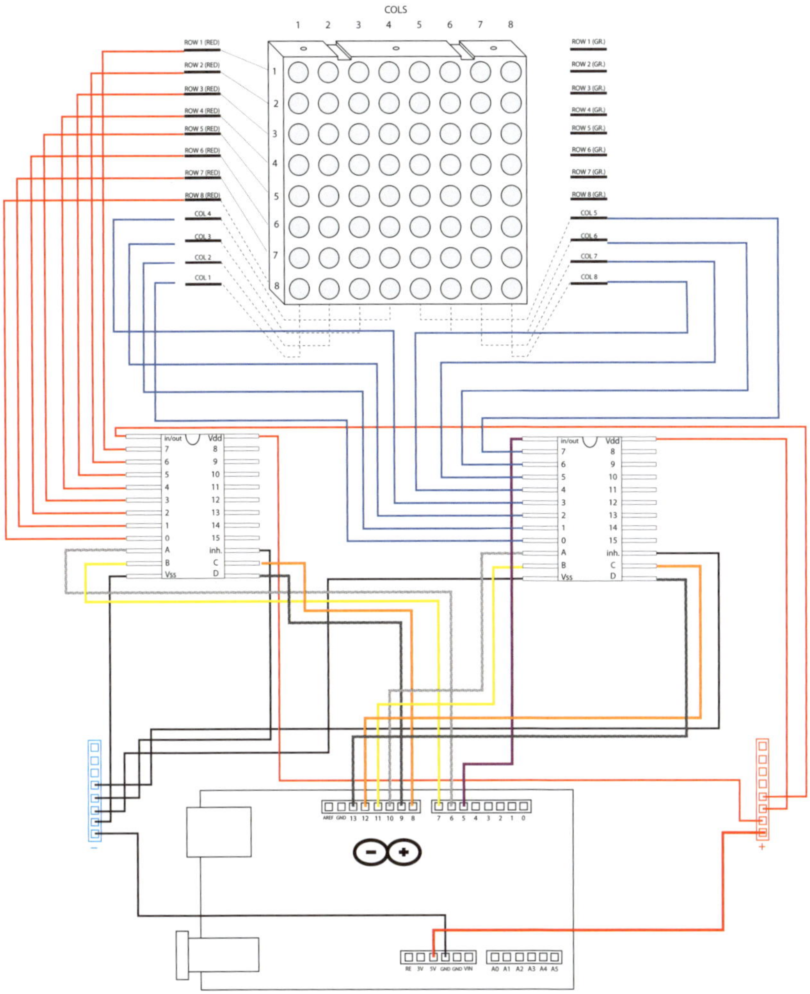

/05

/03-04

A 64-pixel image is extended over the entire block face by means of a carton-like diffuser.

/05

Circuit diagram of Arduino powered prototype.

invisible room

A house in Hong Kong uses optical devices to conceal and digital technology to reconstruct the real
/ Samuel Ko*

Things are not what they seem in this rooftop house. Perception is here actively tricked and mediated by means of digital and analogue devices.

This house occupies the roof of a mid-rise building in a very densely populated area of Hong Kong. It is meant to be as transparent as possible to minimise the obstruction of the view from adjacent buildings. But glass buildings are not entirely transparent and hardly invisible. What is invisible here is the master bedroom on the upper floor that is designed to virtually disappear thanks to a system of mirrors that creates an optical illusion of invisibility. Inside the stealth bedroom, four digital displays designed to substitute windows mediate the perception of the exterior by overlaying a programmed and wilfully reconstructed version of the real. For instance, elements of the view can be deleted or reconstituted to achieve desired effects. Time also could be refashioned with delayed and manipulated video feed, allowing perhaps a view from the digital window into the future by extrapolating from repeating patterns.

/02

/01

/01
View of house from opposite building.

/02
Digital window reconstructs the real – in this demonstration an iphone application seamlessly erases text from a live video image.

/03
Model demonstration of the disappearance of the upper floor room.

/04
Digital windows offer electronically mediated views through a two way mirror in the camouflaged room.

*Guided by Rodolphe el-Khoury

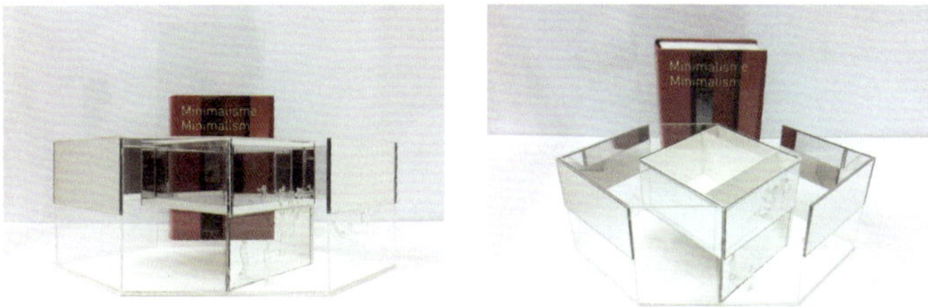

/03

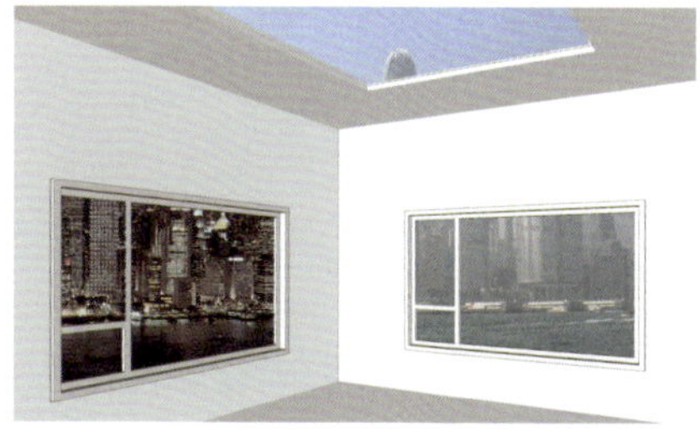

/04

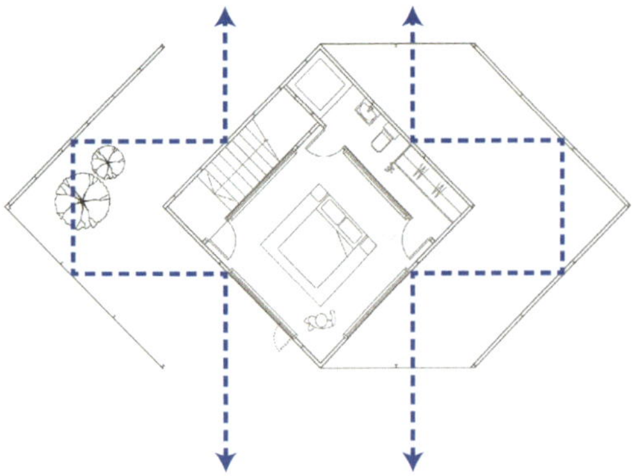

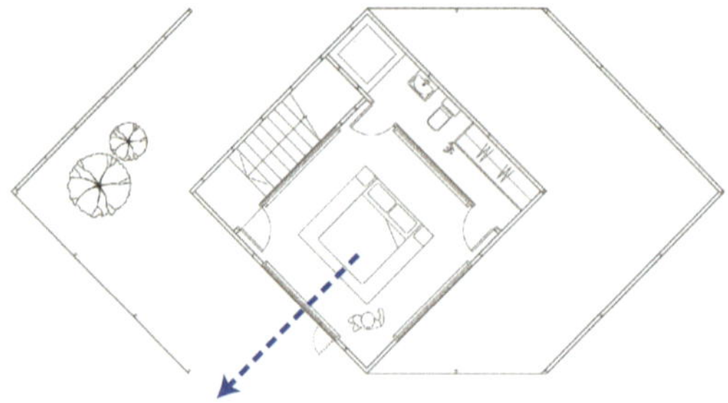

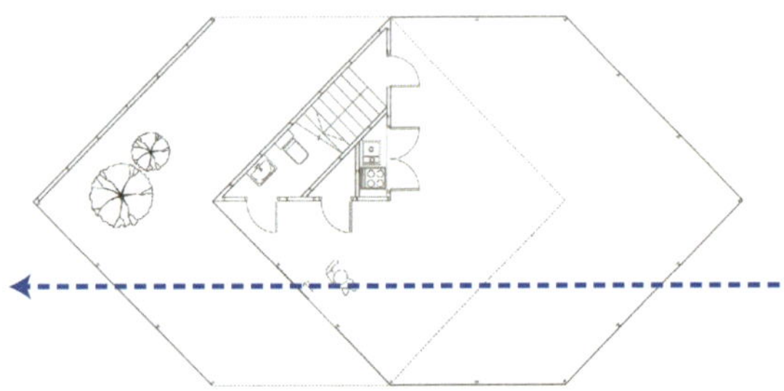

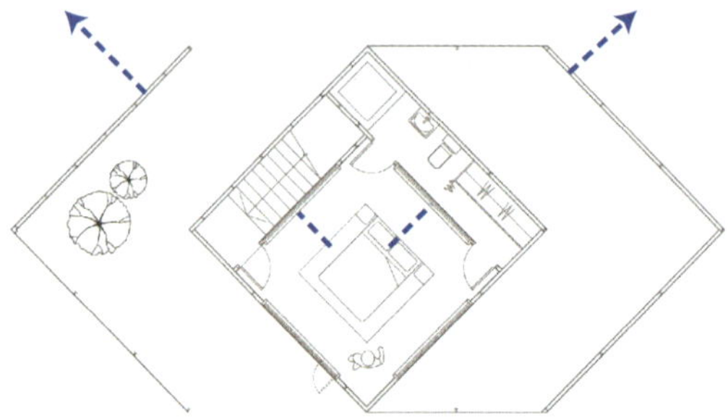

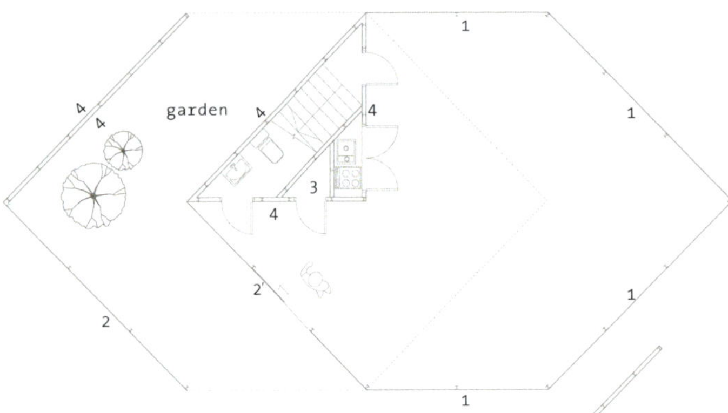

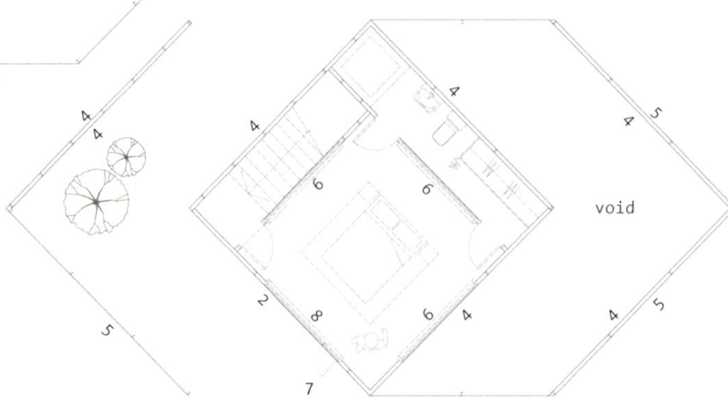

1. glass
2. one way mirror
3. storage
4. mirror
5. one way mirror / samsung OLED monitor
6. digital window
7. openable window
8. analog window

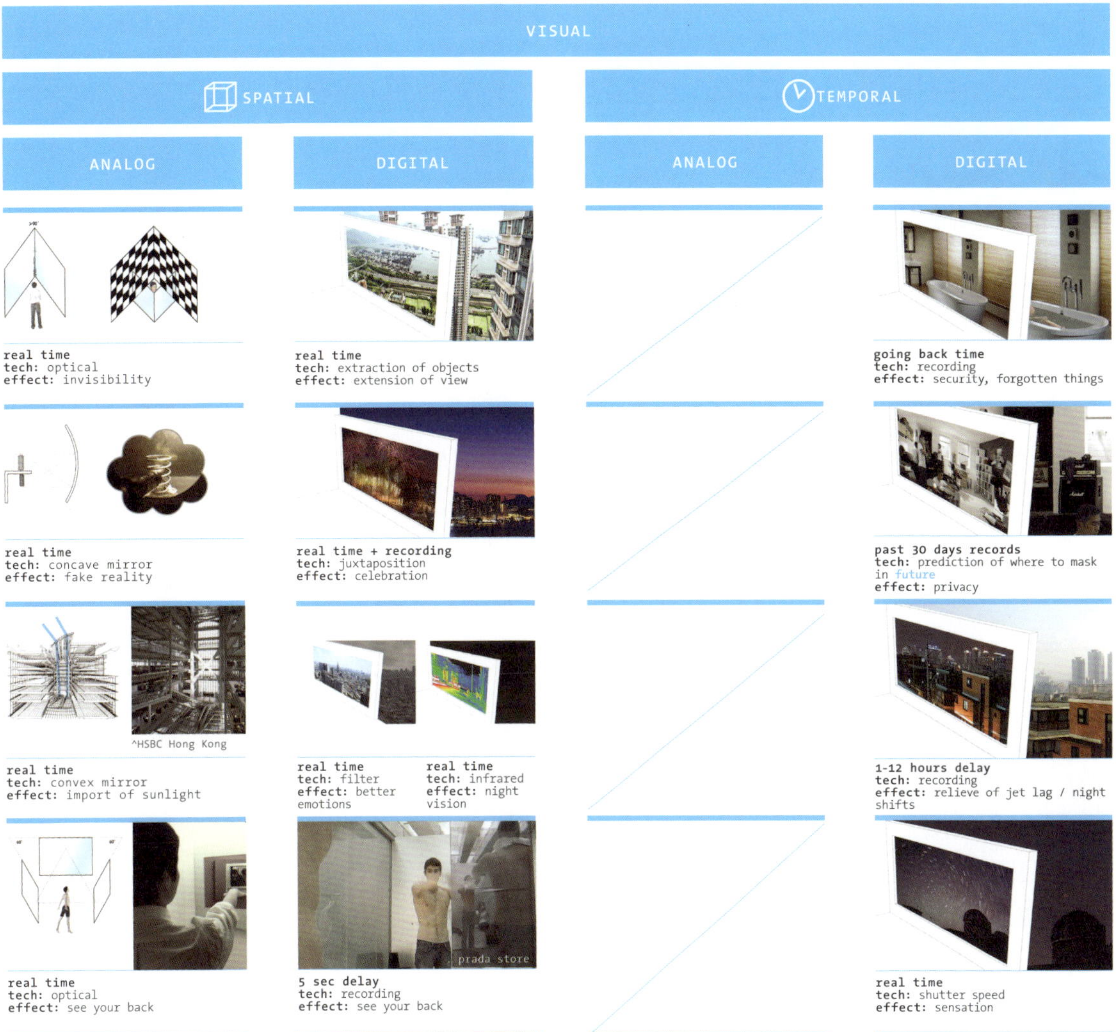

/06

/05

Plan view optical diagrams.

/06

Matrix detailing the possibilities for the reconstitution of the real through electronic mediation.

individuated response

Generic constraints and standardised frameworks give way to customised response, in recognition of a variegated public. Biometric and locative technologies associated with control and surveillance are re-appropriated in favour of subjective needs and desires. The following projects explore new forms of experience in the capacity for environments to dynamically respond to one's actions and attributes. In the public realm, the connectedness is meant to expand an individual's sphere of influence, projecting participants into situations of greater visibility and enhanced sociability. In the domestic environment the responsiveness of the architecture enables different scenarios of inhabitation and lifestyles. The networked and electronically mediated environments exercise a renewed interest in the role that our perceptual and affective faculties have in the construction of collective and individual experience, emphasising the ludic and theatrical, rather than the instrumental or practical facilitation of public participation and private dwelling.

hu-mannequins

Scent emitting mannequins transform a pop-up store / **Studio NMinusOne**

Hu-mannequins is a pop-up store proposal for a fashion designer. The proposal attempts to enrich the experience of viewing by engaging the olfactory sense.

The installation is composed of scent-emitting mannequins that release a variety of smells into the environment throughout the day. Seats are placed within the store to harness the scent of sitting individuals. The scent of those occupying the seats is transported through a system of ducts, where the smell is released into the air via perforated mannequins. The person occupying the seat metaphysically wears the clothing through smell. The mannequin, typically a lifeless object, is now animated by human qualities through the production of smells – becoming a Hu-mannequin.

While sitting on the furniture, a person literally inhabits the clothing, giving an outfit that person's unique smell. This sets up a new dynamic. Unlike the act of physically trying on a suit in a dressing room, partners, friends, and strangers can visualise an outfit by smelling the virtual wearer's body on the clothing.

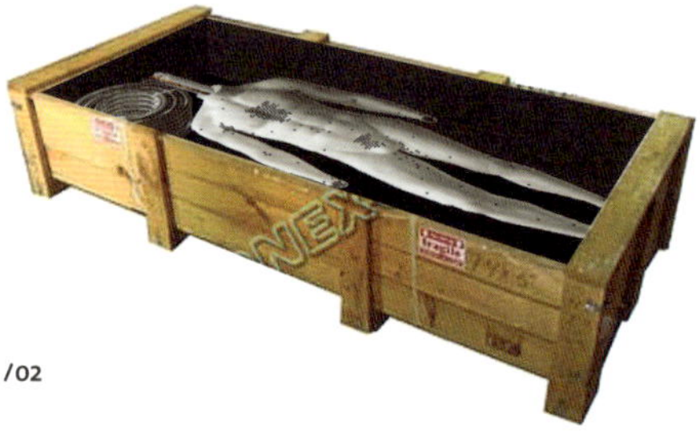

/02

/01

/01
An innocent bystander wearing the suit.

/02
The Hu-mannequin ready for shipping.

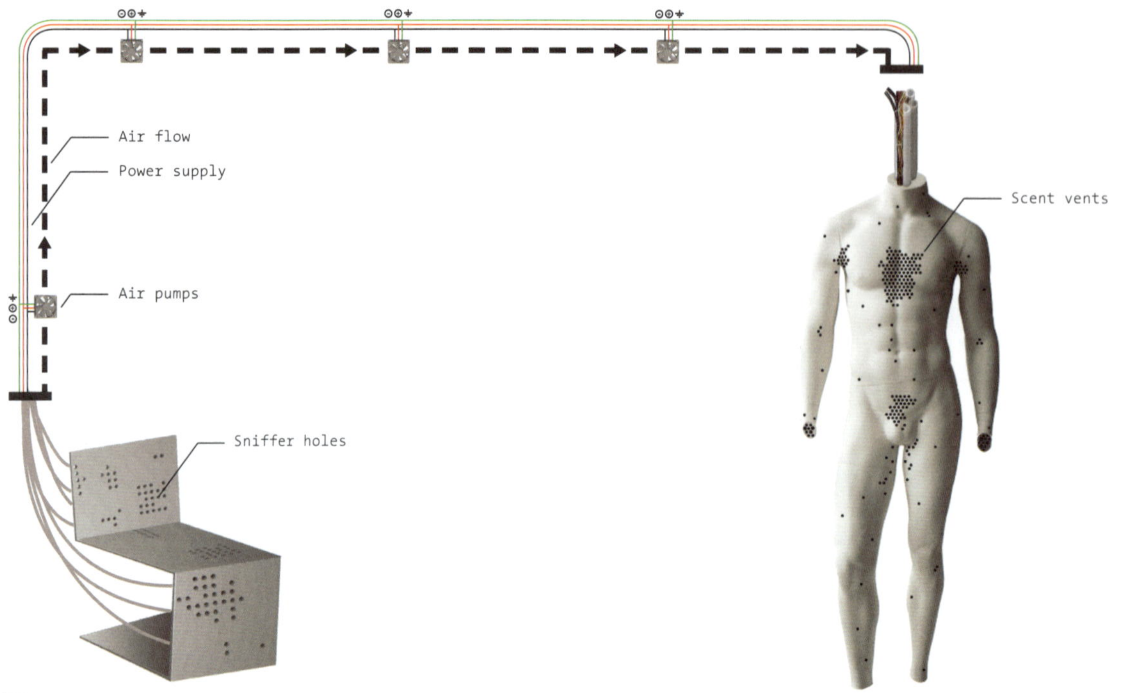

/03

/03
How it works: air is sucked in through the perforations in the chair, relayed through ducts to the mannequin's body where scents are released in their corresponding areas.

/04
Exterior of pop-up store underneath New York City's High-Line.

/04

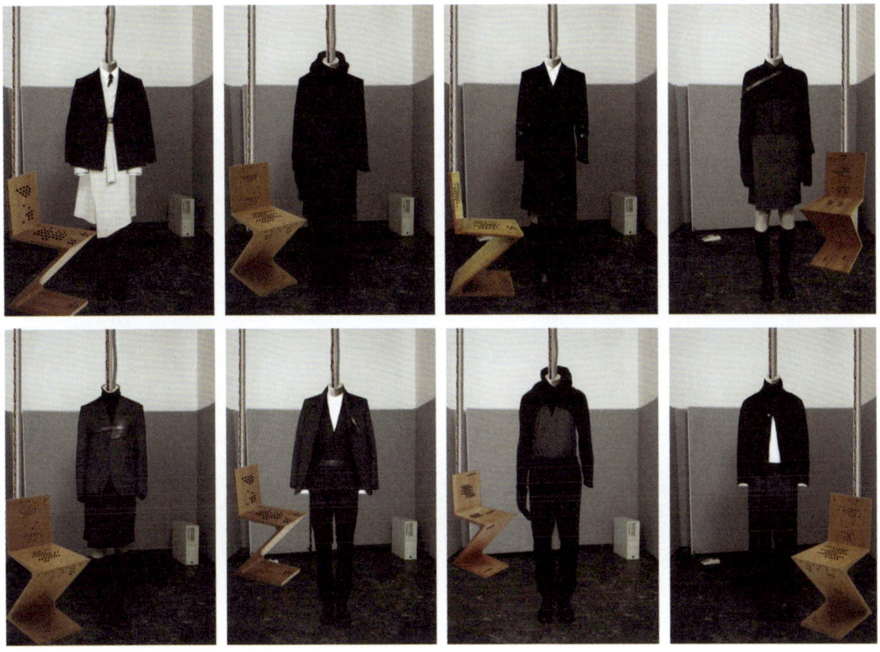

spot shade

Vertical shades modulate the light by tracking the movement of the sun / **Melissa Lui***

Spot Shade is an interactive shading device that constantly tracks the location of sun in the sky with a light-seeking mechanism, regulating and projecting exterior light into the interior space to provide desirable, natural lighting conditions.

Spot Shade is installed behind a glass wall and is digitally enhanced to regulate natural light, creating desirable lighting conditions for a particular reading spot inside. Uniquely customised for a specific internal location and actuated if that location is occupied, the project examines the possibility of using digital technology to achieve the ultimate customisation of the spatial experience.

The prototype consists of five hanging sculptural bars installed behind a glass wall. Each bar is individually controlled by a servo motor connected to an Arduino board. The device is programmed so that, depending on the registered location of the sun, the bars are rotated to ward off the direct sunlight from a particular reading spot while letting in ambient light. Five light sensors are installed within the gap between the shading device and the glass wall. Computing the differential of light intensity detected by these sensors, the Arduino board pin-points the direction of the sunlight and actuates the shading device accordingly. To make the digitally enhanced experience a hyper-customised one, the device is activated only when the reading spot is actually occupied. A single chair placed at the reading spot is equipped with a pressure sensor that detects when someone is sitting in it. This information is communicated to the Arduino board to activate the shading device.

/01
View of shading system from outside.

/02
Diagram showing local blocking of direct sunlight.

*Guided by Rodolphe el-Khoury and Nashid Nabian

/01

/02

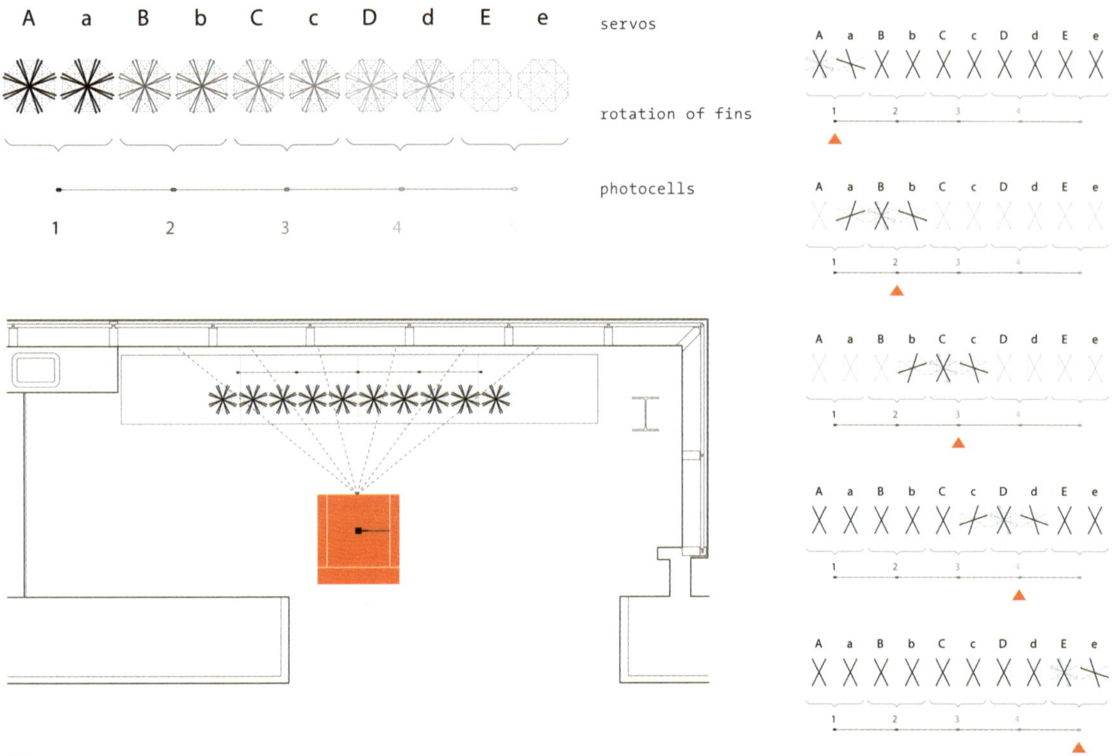

/03

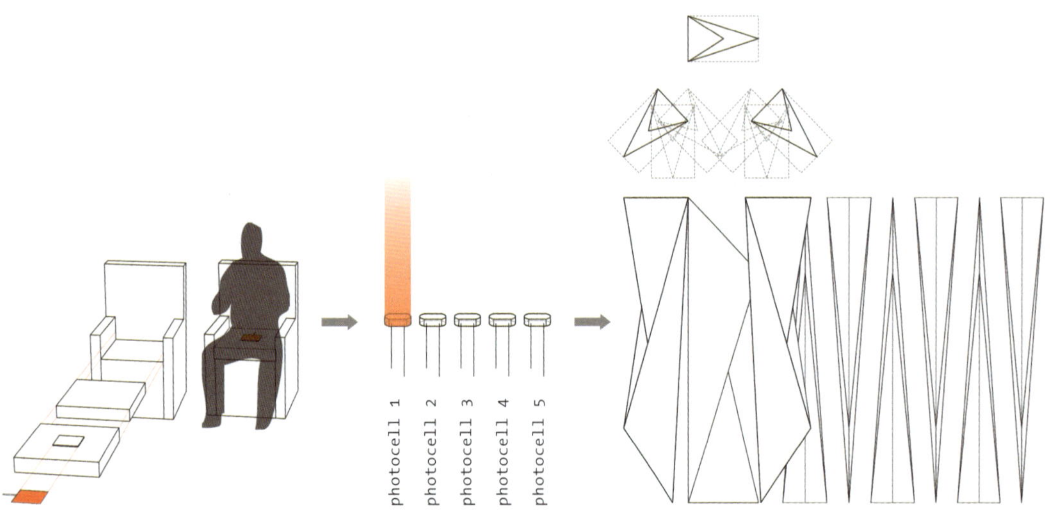

^ pressure sensor detecting presence in chair

^ rotating fins

/04

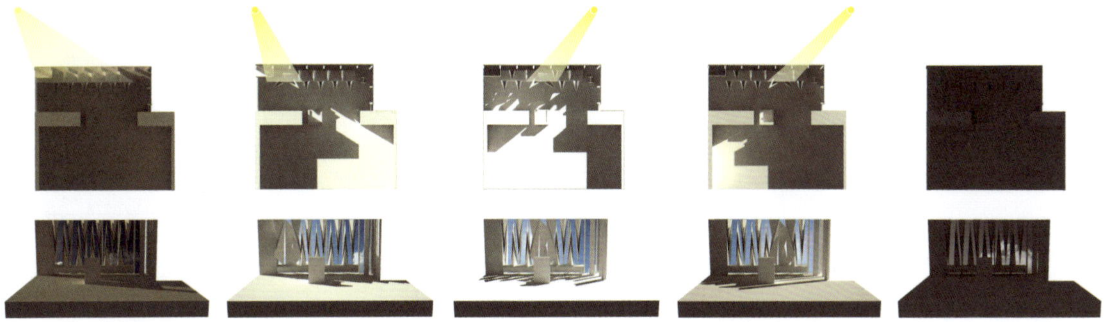

/05

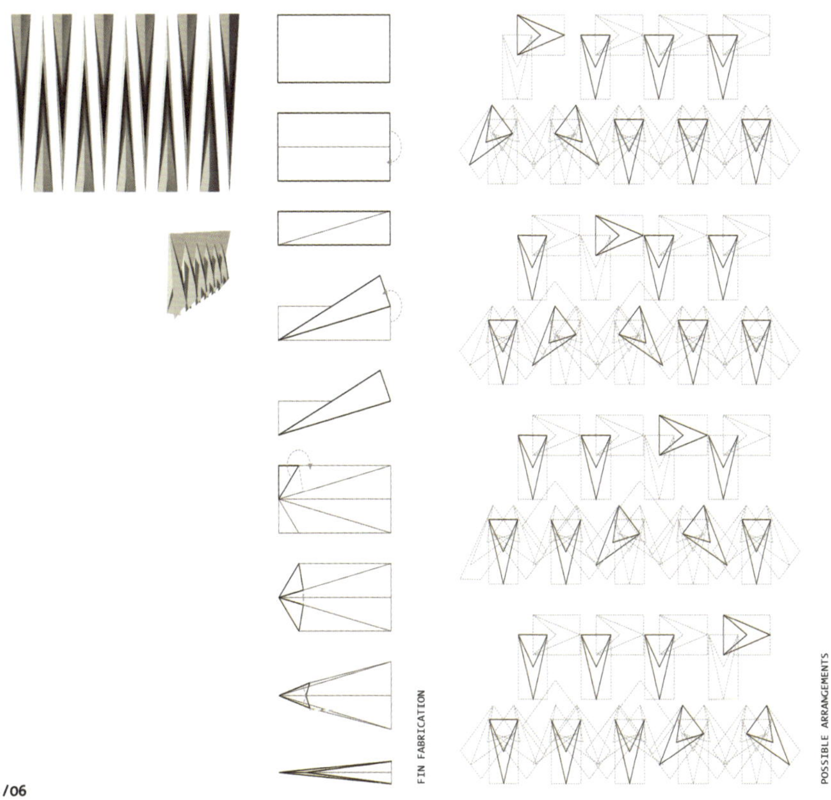

/06

/03
Floor plan of reading alcove in the library at the Daniels Faculty of Architecture, Landscape and Design and sequence for screen activation.

/04
Screen as viewed from exterior of the building.

/05
Movement of fins over the course of a 12 hour period from 6am to 6pm.

/06
Plan view of rotating shades.

auratic chamber

This environment uses light and colour to become an extension of one's psychological state / **Jimmy Tran***

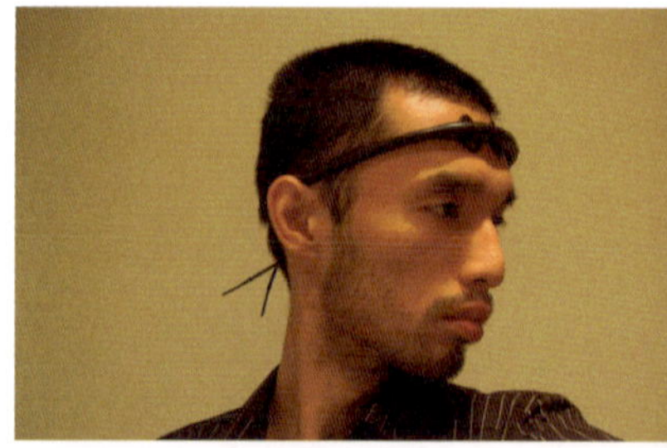

Auratic Chamber creates a variety of spatial experiences without the usual arrangements of walls, floors or ceilings. Instead, light is used to create experiences which vary depending on the inhabitant's psychological state.

The lighting of the room interacts with the inhabitant's physiological and psychological state – measured by a neural impulse sensor worn by the inhabitant. The room creates a halo of light according to the person's behaviour and temperament. The inhabitant can consciously interact with this halo, treating it as an extra limb of his or her body, or an extension of his or her psyche.

As more people enter the room, each person creates their own halo. Each person's halo follows them as they traverse the space, providing a new form of biofeedback and self-expression. The halo of light and colour becomes a new form of body language, giving each person a heightened state of self-awareness, and sending visual cues to others about each person's emotional and psychological state.

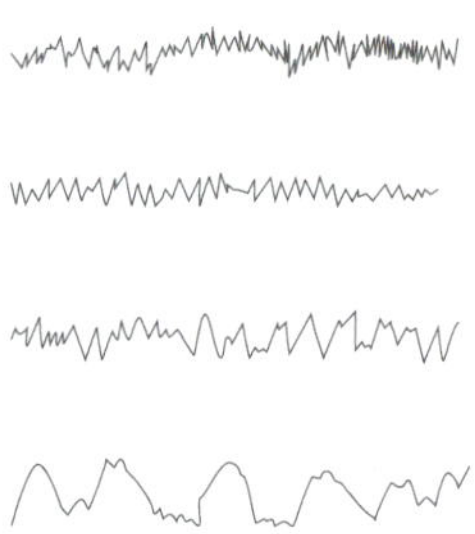

Beta 15-30Hz
Awake, Alert, Consciousness

Alpha 9-14Hz
Relaxed, Calm, Meditation Creative Visualisation

Theta 4-8Hz
Deep Relaxation and Meditation Problem Solving

Delta 1-3Hz
Deep, Dreamless Sleep

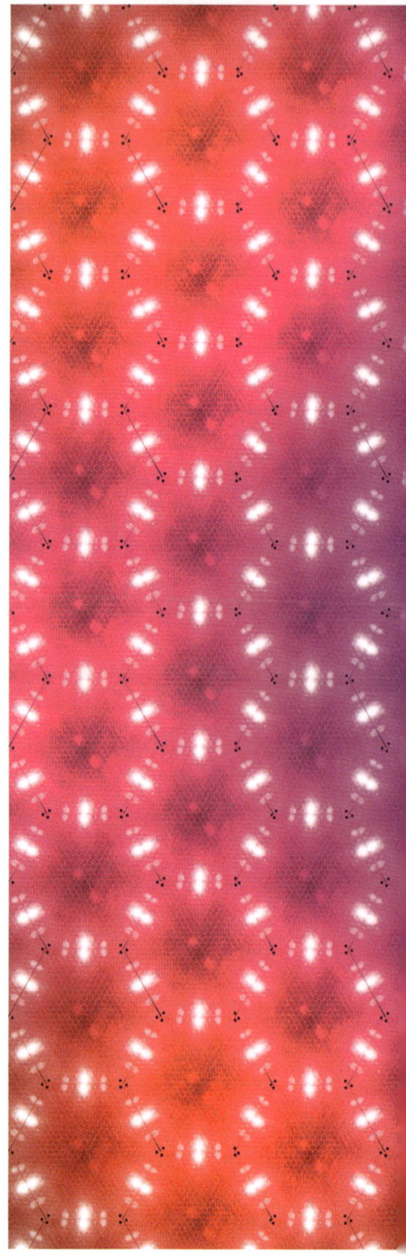

/01
The Auratic Chamber consists of interlocking tiles that line the interior surfaces of a space.

*Supervised by Christos Marcopoulos

/01

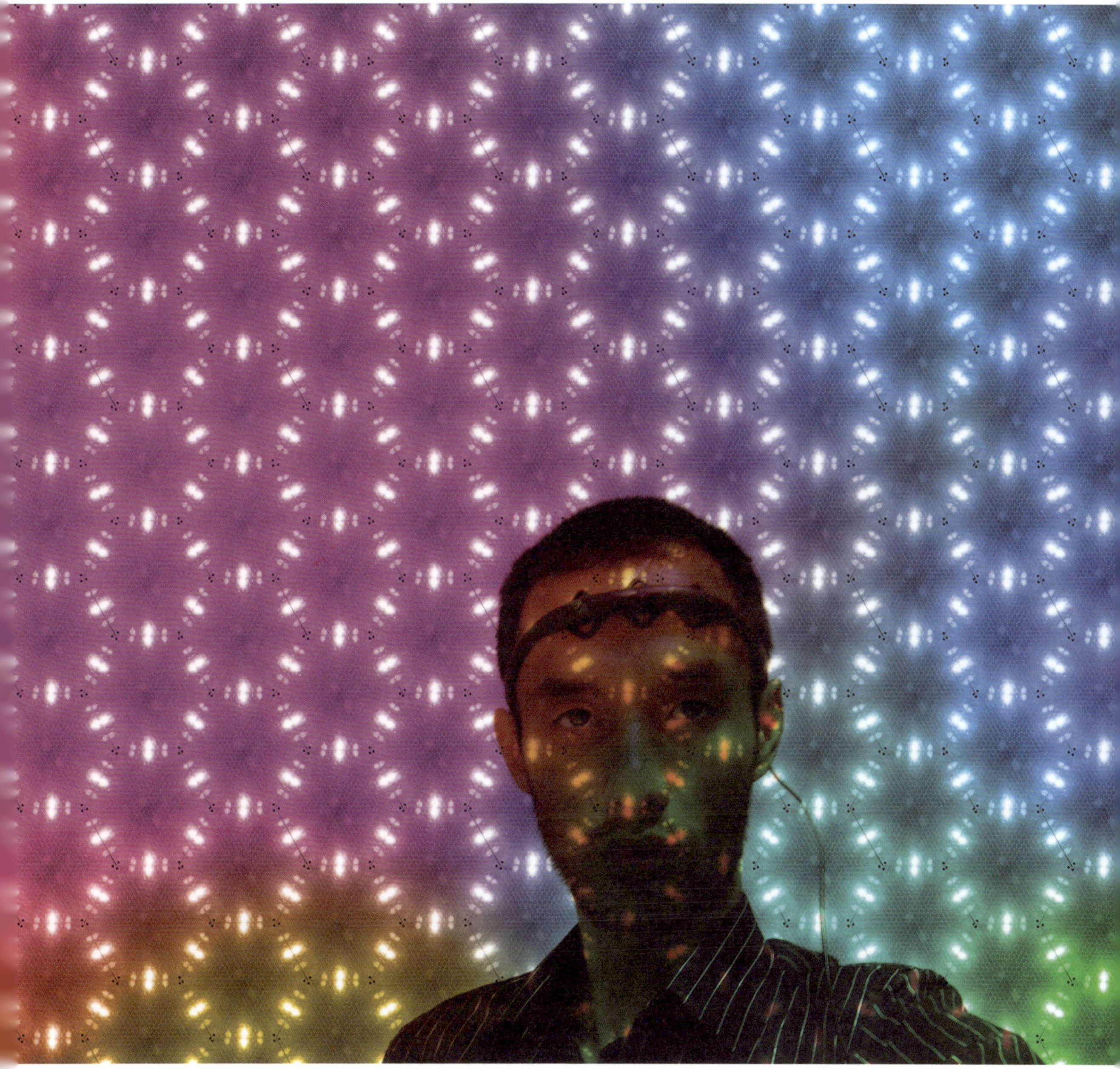

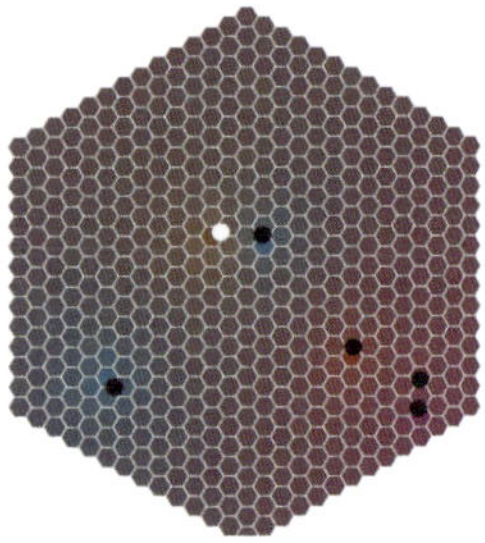

PERSON #6

Emotions

Calm — Excited
Bored — Engaged
Meditation
Frustration

Expressions

Clenching

Blink
Smiling
Laughing
Smirking

Brainwave Frequency

Beta 15-30Hz
Awake, Alert, Consciousness

Alpha 9-14Hz
Relaxed, Calm, Meditation
Creative Visualization

Theta 4-8Hz
Deep Relaxation and Meditation
Problem Solving

Delta 1-3Hz
Deep, Dreamless Sleep

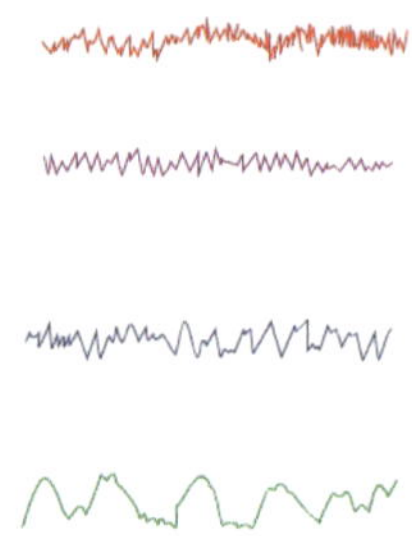

Conscious Commands

Lift	Left	Disappear
Drop	Right	
Push	Clockwise	
Pull	Counter-Cw	

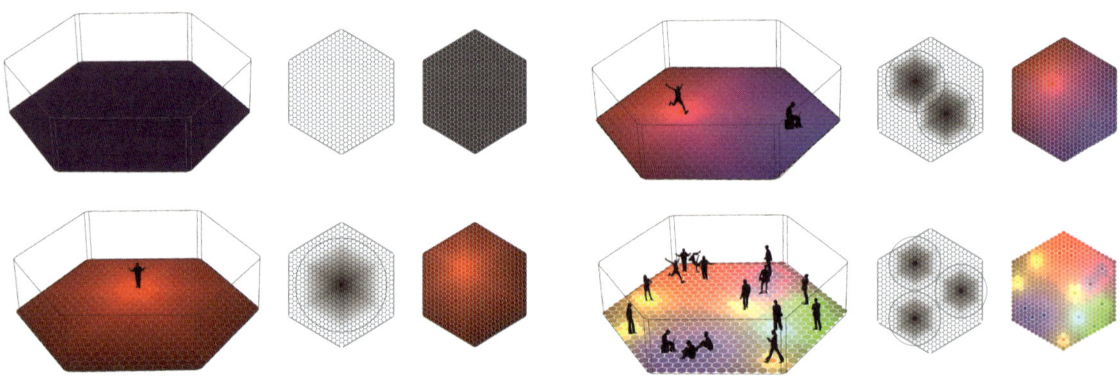

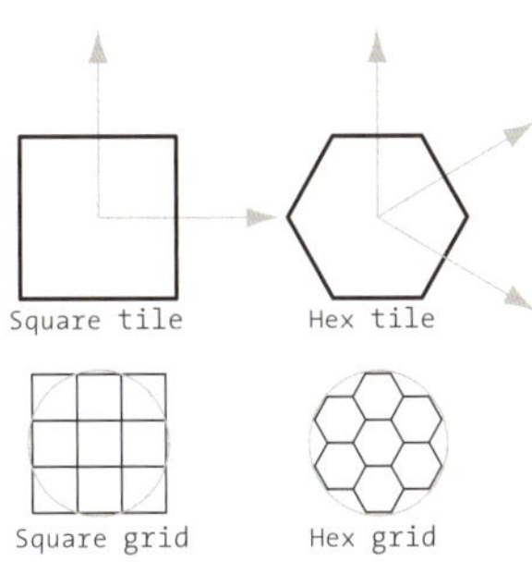

Hex tiles offer 3 axis movement between tiles and a radial grid

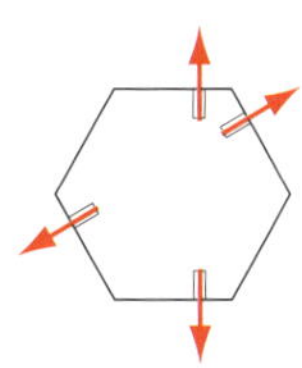

Tile offers four points of connection to assemble a hex grid

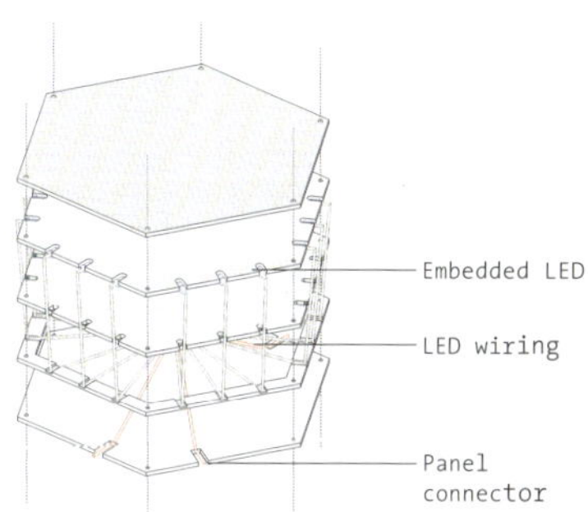

/03
This sequence shows the space under different social circumstances. Some people light up a room, while others have the opposite effect.

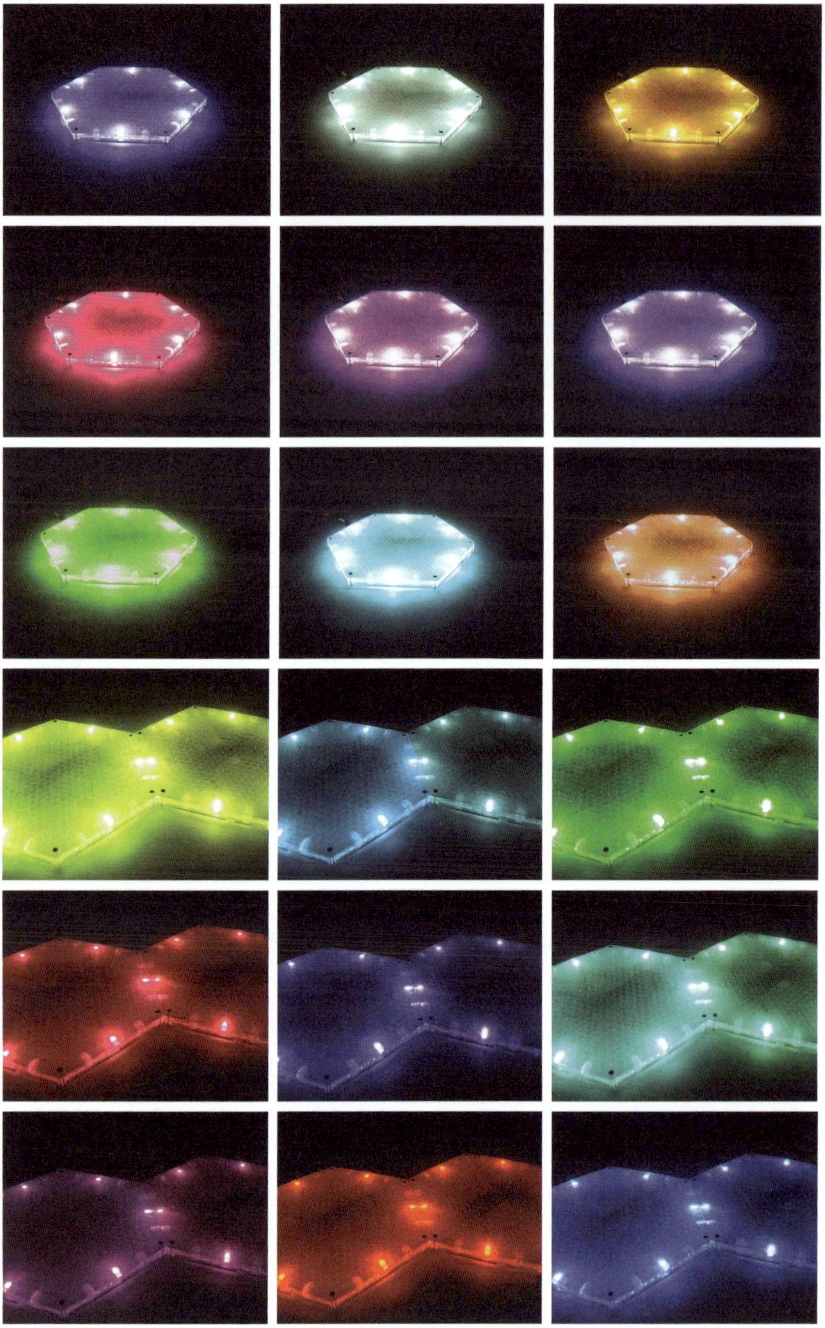

Prototypes of individual and interlocking tiles.

scentisizer

Scent Engine and Tangible Interface for Composing and Performing Scentscapes / **Khoury Levit Fong with Nashid Nabian**

Scentisizer allows for the orchestration and delivery of complex fragrances by giving precise control over the constitution and diffusion of scent accords in dynamic compositions.

A tangible interface combines haptic and graphic features for managing an array of sixty-four scent dispensers that maps a wide-ranging olfactory field. Users can manipulate individual scent containers, much like organ stops, to compose and time an accord in a direct and intuitive fashion. They can also control them with greater complexity and precision by means of a digital graphic user interface that allows for multi-channelled sequencing and modulation of scent tones and dynamics. The analogue controls provide the haptic means for tangibly sculpting and visualising olfactory phenomena while the digital features add functionality and web connectivity, allowing Scentisizer to perform digitised olfactory scores stored locally or accessed wirelessly.

With the haptic interface users can control the behaviour of their scent-compositions over a prescribed period – programmed by manipulating the array of dispensers. The slightest elevation from the default low position activates a dispenser while varying elevations trigger different pre-programmed diffusion cycles. Once the cycle is launched, the motorised tubes are automatically animated, sinking gradually into their default rest position while transcoding the status of the unfolding olfactory phenomenon into a topographic image at any given time to build a tangible and instantly graspable 3-d diagram of the evolving scent structure.

The graphic user interface enables greater control over the scent-dispensing array, providing user-adjustable parameters for shaping the dynamics of the individual constituents of the fragrance and the overall effect over time.

/01
Scentisizer in 'perform' mode at InterAccess, Toronto.

/02
Demonstration and prototype of haptic user interface.

/03
Scent mapping. The scents assigned to each dispenser are mapped according to a variation on the Fragrance Wheel, a taxonomy developed by the fragrance industry for a user-friendly classification of scents in distinct but related families: Floral, Fresh, Oriental, Woody, and Aromatic Fougère.

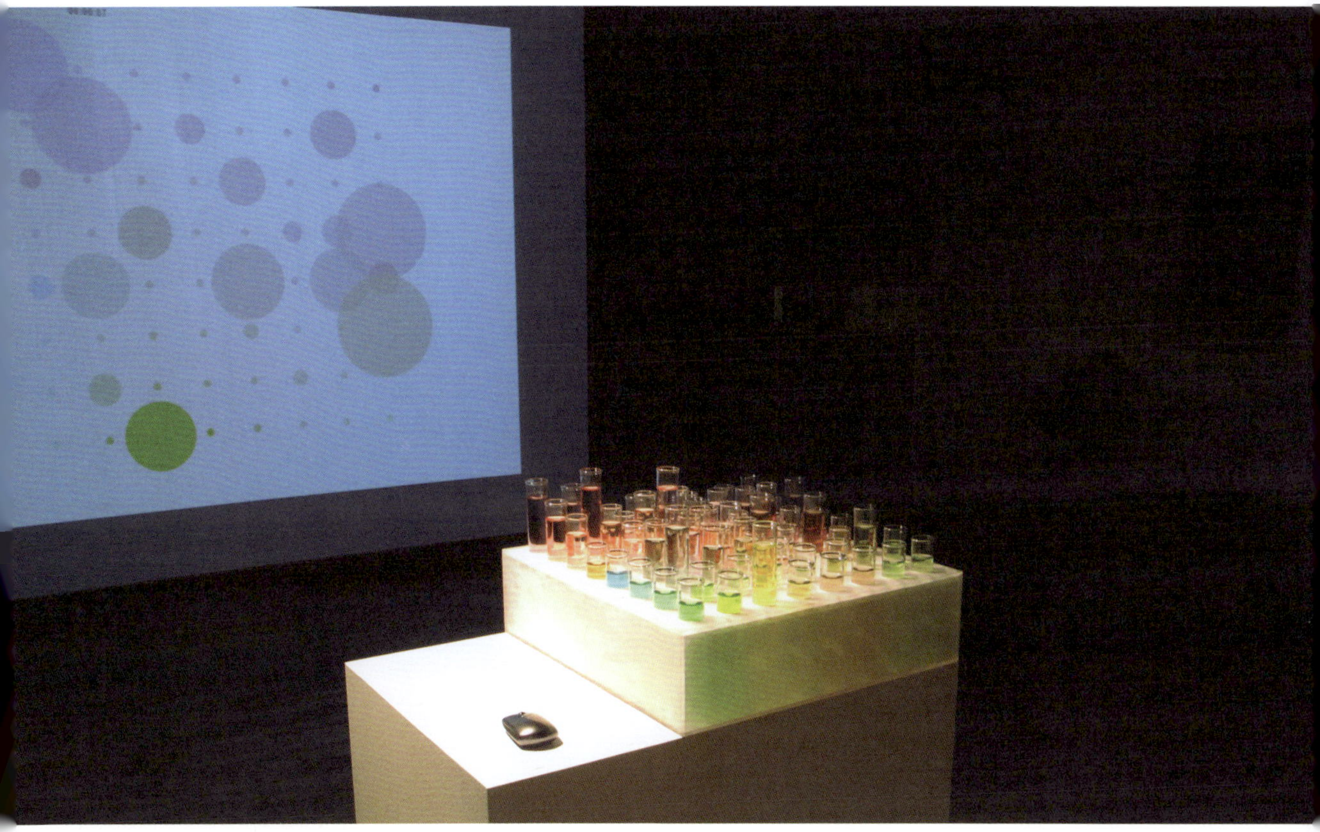

/01

/02

/03

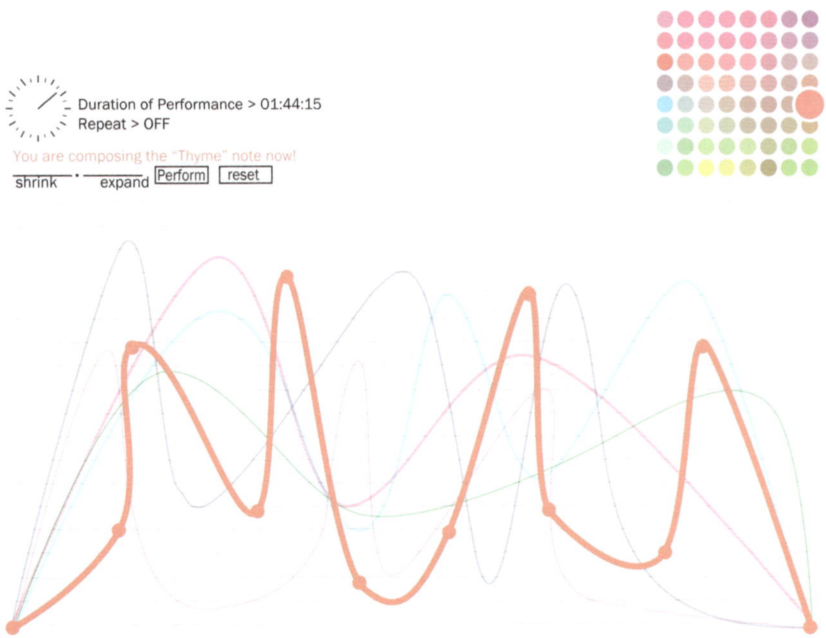

/04

/04
Graphic user interface for composing and performing scentsapes.

/05
Circuit diagram of haptic user interface.

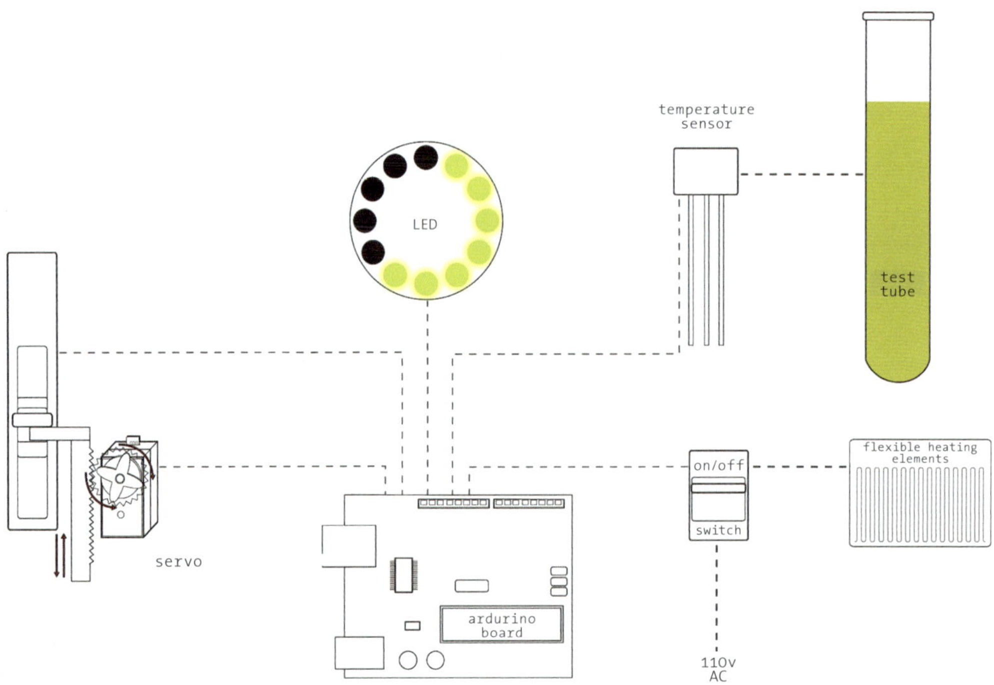

/05

chameleon lounge

Sensor equipped chairs imitate the colour of users' clothing and objects with which they come in contact / **Khoury Levit Fong with Nashid Nabian**

Chameleon Lounge is a collection of plastic moulded chairs that have been electronically enhanced to respond to colour. Embedded technology allows each chair to pick up the colour of the person who uses it and to imitate it in a chameleon-like fashion.

Chameleon Lounge belongs to a series of projects by Khoury Levit Fong which aligns architecture with information technology to empower the sensate body and its spatio-temporal envelope as the intimate locus of social connection. It explores, in a modest application, the capacity for environments to respond precisely and intimately to a subject's actions and characteristics. The individuated connectedness seemingly amplifies the subject's charismatic aura, thus expanding his or her sphere of influence in public.

The electronics are comprised of a colour sensor, an Arduino microcontroller and high-intensity RGB LEDs. The Arduino is programmable for a variety of scenarios. In the New Orleans version, the LEDs gradually turn off when the chair is not in use and the embedded sensor does not pick up a new colour. In Casablanca, the LEDs are programmed to cycle through a sequence of colours when the chairs are unoccupied, simulating spontaneous swarm behaviour when left undisturbed.

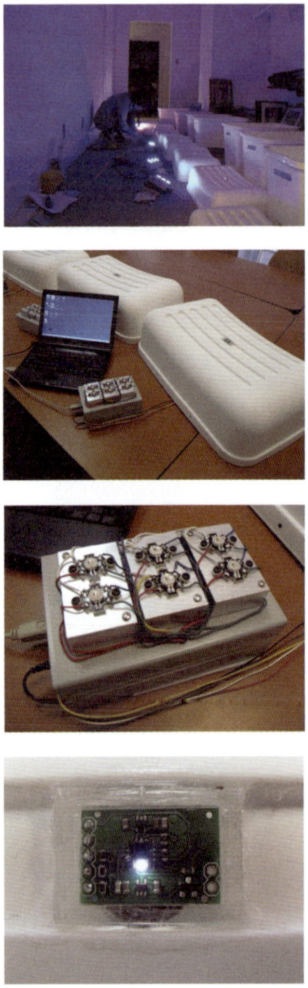

/01

/01
Electronic components integrated into IKEA sourced stool.

/02
Installation in operation at New Orleans AIA DesCours Festival 2008.

/02

Installation in operation at Casablanca art festival in 2009.

appendix

authors

Rodolphe el-Khoury

Rodolphe el-Khoury is Associate Professor at the University of Toronto's Daniels Faculty and partner in the design firm Khoury Levit Fong. He is the author of numerous critically acclaimed books in architectural history and theory and a regular contributor to professional and academic journals. His books include *Monolithic Architecture*, *Architecture: In Fashion*, *Shaping the City: Studies in History, Theory and Urban Design*, and *See Through Ledoux: Architecture, Theatre and the Pursuit of Transparency*. He has received several awards and international recognition for his design work as principal at Office dA, ReK Productions and currently at KLF. El-Khoury is the co-director of RAD, a research unit at the University of Toronto dedicated to architectural applications for advanced information technology, aiming for enhanced responsiveness and sustainability in the built environment.

Carol Moukheiber

Carol Moukheiber is Assistant Professor and Director of Urban Design at the University of Toronto's Daniels Faculty of Architecture, Landscape, and Design. She is also co-founder and partner in the architecture practice Studio NMinusOne (n-1). At the domestic scale, her work has focused on the home as an immersive environment – one capable of generating new physical and emotional experiences – through the enhancement or recalibration of its infrastructure, or set of programmes. She is the co-editor of *Wild Wild Urbanism, Redesigning California* (CCA 2006). Her design work has been acquired by the San Francisco Museum of Modern Art, and published widely in academic and mainstream media, including *The New York Times Magazine* and *Praxis Journal of Architecture*. She is the recipient of a Canada Foundation for Innovation grant providing support for the newly launched RAD. Studio NMinusOne has been selected by the New York Architectural League for their Emerging Voices lecture series, 2012.

Christos Marcopoulos

Christos Marcopoulos is Assistant Professor at the University of Toronto's Daniels Faculty of Architecture, Landscape, and Design. He is also co-founder and partner in the architecture practice Studio NMinusOne (n-1). Marcopoulos has extensive professional working experience having worked in the offices of OMA, (Office for Metropolitan Architecture, Rotterdam), led by Rem Koolhaas, and Skidmore, Owings and Merrill, San Francisco. His technical expertise has been instrumental in the development of Studio NMinusOne's projects. His work on the domestic environment has been acquired by the San Francisco Museum of Modern Art. He is the co-editor of *Wild Wild Urbanism, Redesigning California* (CCA 2006). His built work has been published widely in academic and mainstream media, including *The New York Times Magazine* and *Praxis Journal of Architecture*. Studio NMinusOne has been selected by the New York Architectural League for their Emerging Voices lecture series, 2012.

credits

Studio NMinusOne Projects Information:

–

"IM Blanky"
2011 / Product Prototype
Exhibition: Stitches, curated by Larry Wayne Richards, WORKShop Gallery, Toronto
Principals in Charge: Christos Marcopoulos and Carol Moukheiber in collaboration with Rodolphe el-Khoury
Team: Valentina Mele, Sebastian Savone, Jonah Ross-Marrs, Yie Ping See, Samar Sabie, Dina Sabie

–

"Farnsworth Curtain"
2011 / Interior Design Show Installation Proposal
Location: Toronto
Principals in Charge: Christos Marcopoulos and Carol Moukheiber with Rodolphe el-Khoury
Team: Min Woo Kim, Andria Fong

–

"Hu-mannequins"
2010
Competition: Architizer, House of Fashion for Siki Im, New York
Principals in Charge: Christos Marcopoulos and Carol Moukheiber
Team: Valentina Mele, Michael Spatafora

–

"Out-House"
2009
Exhibition: Out of Water, curated by Liat Margolis and Aziza Chaouni, LWR Gallery, Toronto
Principals in Charge: Christos Marcopoulos and Carol Moukheiber in collaboration with Don Shillingburg
Landscape Consultant: Rob Wright

–

"RGB Garden"
2009
Competition: The 11th International Metis Garden Festival, Quebec
Principals in Charge: Christos Marcopoulos, Carol Moukheiber in collaboration with Nashid Nabian
Team: Melissa Lui, Valentina Mele, Cleo Buster

–

"Dual State Space"
2008 / Product Prototype
Principals in Charge: Christos Marcopoulos and Carol Moukheiber
Team: Michael de Jong, Melissa Lui

–

"Digital Window"
2008 / Product Prototype
Principals in Charge: Christos Marcopoulos and Carol Moukheiber

–

"Farnsworth Wall"
2008 / Product Prototype
Principals in Charge: Christos Marcopoulos and Carol Moukheiber in collaboration with Mani Mani

–

"Sky House"
2004
Principals in Charge: Christos Marcopoulos and Carol Moukheiber

–

"Invisible House"
2004 / Product Prototype
Principals in Charge: Christos Marcopoulos and Carol Moukheiber
Team: Patrick Tierney, Michael de Jong

KLF Projects Information:

–

"Rule"
2011 / Installation
Location: MIT, Cambridge
Location: MIT School of Architecture
Principals in Charge: Rodolphe el-Khoury and Robert Levit
Team: Zachariah Glennon, Lindsay Hochman, Renee Leung, Ultan Byrne

–

"Nature 2.0"
2011 / Installation
Location: Harborfront Gallery, Toronto, Ontario
Principals in Charge: Rodolphe el-Khoury and Robert Levit
Team: Nariman Mousavi and Zachariah Glennon, Faisal Bashir and Irina Solop

–

"Tunnel of Light"
2011 / Interior Design Show Installation Proposal
Location: Toronto, Ontario
Principals in Charge: Rodolphe el-Khoury and Robert Levit
Team: Zachariah Glennon, Nariman Mousavi

–

"Superblock"
2011 / Product Prototype
Principal in Charge: Rodolphe el-Khoury in collaboration with Nashid Nabian
Team: Zack Glennon, Nariman Moussavi

–

"Blink"
2010
Competition: Land Art Generator Initiative
Principals in Charge: Rodolphe el-Khoury and Robert Levit
Team: Lindsay Hochman, Khaled Al-Nasser, Melissa Lui

–

"Monuments and Bits"
2009 / Exhibition
Location: Eric Arthur Gallery, Toronto; CMKY Gallery, Ann Arbor
Principals in Charge: Rodolphe el-Khoury and Robert Levit
Team: James Dixon, Mike Fung, and Renée Leung

–

"Linking Shade"
2009
Competition: Gimme Shelter, Phoenix
Awards: Finalist
Principals in Charge: Rodolphe el-Khoury and Robert Levit in Collaoration with Kuth Ranieri Architects
Team: Mike Fung

–

"MyCity"
2009 / Installation
Location: Convenience Gallery, Toronto
Principal in Charge: Rodolphe el-Khoury
Programming: Mark Jones (Emergent Software)
Team: James Dixon

–

"AlephTO"
2009 / Installation
Location: Nuit Blanche, Toronto
Principal in Charge: Rodolphe el-Khoury in collaboration with Nashid Nabian
Team: Michael Fung

–

"Scentisizer"
2009 / Installation
Location: InterAccess, Toronto
Principal in Charge: Rodolphe el-Khoury in collaboration with Nashid Nabian
Team: Ali Yarbakht, Steven Avis, Mahan Javadi

acknowledgments

The projects documented in this book have benefitted tremendously from the resources, ideas and research at the academic programs, cultural institutions, and professional offices where they developed; we thank the following for their support: California College of the Arts, Convenience Gallery, AIA New Orleans, HarbourFront Center, InterAccess, Khoury Levit Fong, Massachusetts Institute of Technology, Studio NMinusOne, University of Hong Kong, University of Michigan, University of Toronto and WORKshop.

We especially thank the University of Toronto's Daniels Faculty, the MIT Department of Architecture, and the University of Hong Kong's Faculty of Architecture for supporting this publication.

Among several peers, colleagues and friends who have contributed critical insights and expert knowledge to the work we especially recognize and thank George Baird, Tom Bessai, Adrian Blackwell, Brian Boigon, Steven Fong, Jordan Geiger, Mark Goulthorpe, Jason Halter, Anita Matusevics-Halter, Mitchell Joachim, Sheila Kennedy, Ted Kesik, Robert Levit, David Lieberman, Nashid Nabian, Andrew Payne, Larry Wayne Richards, Elise Shelley, John Shnier, Jonathan Solomon, Richard Sommer, Nader Tehrani, Tom Verbes, Mason White, Shane Williamson and Meejin Yoon.

We thank the book's production team for flawlessly orchestrating the publication of the work: Melissa Lui, Valentina Mele, Leo Malinow, Kit Maude, Oscar Riera Ojeda, Alejandra Román and Ya'el Santopinto.

We are also grateful to Diane and Victor Mele for supporting this project.

–

"Chameleon Lounge"
2008 / Installation
Location: DesCours Arts Festival, New Orleans, Louisiana
Principal in Charge: Rodolphe el-Khoury in collaboration with Nashid Nabian
Team: Robert Levit, Nashid Nabian, Mike Fung, James Dixon

–

"Window 2.0"
2008 / Installation
Location: LWR Gallery, Toronto Principal in Charge: Rodolphe el-Khoury
Team: James Dixon

–

"MOCAPE"
2007 / Competition
Location: Shenzhen
Awards: Competition Finalist (First of four/initial competition) AIA Merit award for unbuit architecture
Principals in Charge: el-Khoury, Levit, Fong
Landscape Architects: Alissa + Peter North, NDO
Team: Talal Rahmeh, Shan Li, Raj Patel, Adrian Phiffer, Luke Stern, Azia Dhamani, Yusef Frasier, Mike Fung, Nick Maroulis

–

"Stratford Square"
2006
Competition: Market Square, Stratford
Location: Stratford, Ontario
Awards: Honourable Mention in Open Competition
Principal in Charge: Rodolphe el-Khoury
Team: Lingchen Liu and Chenglong Wang

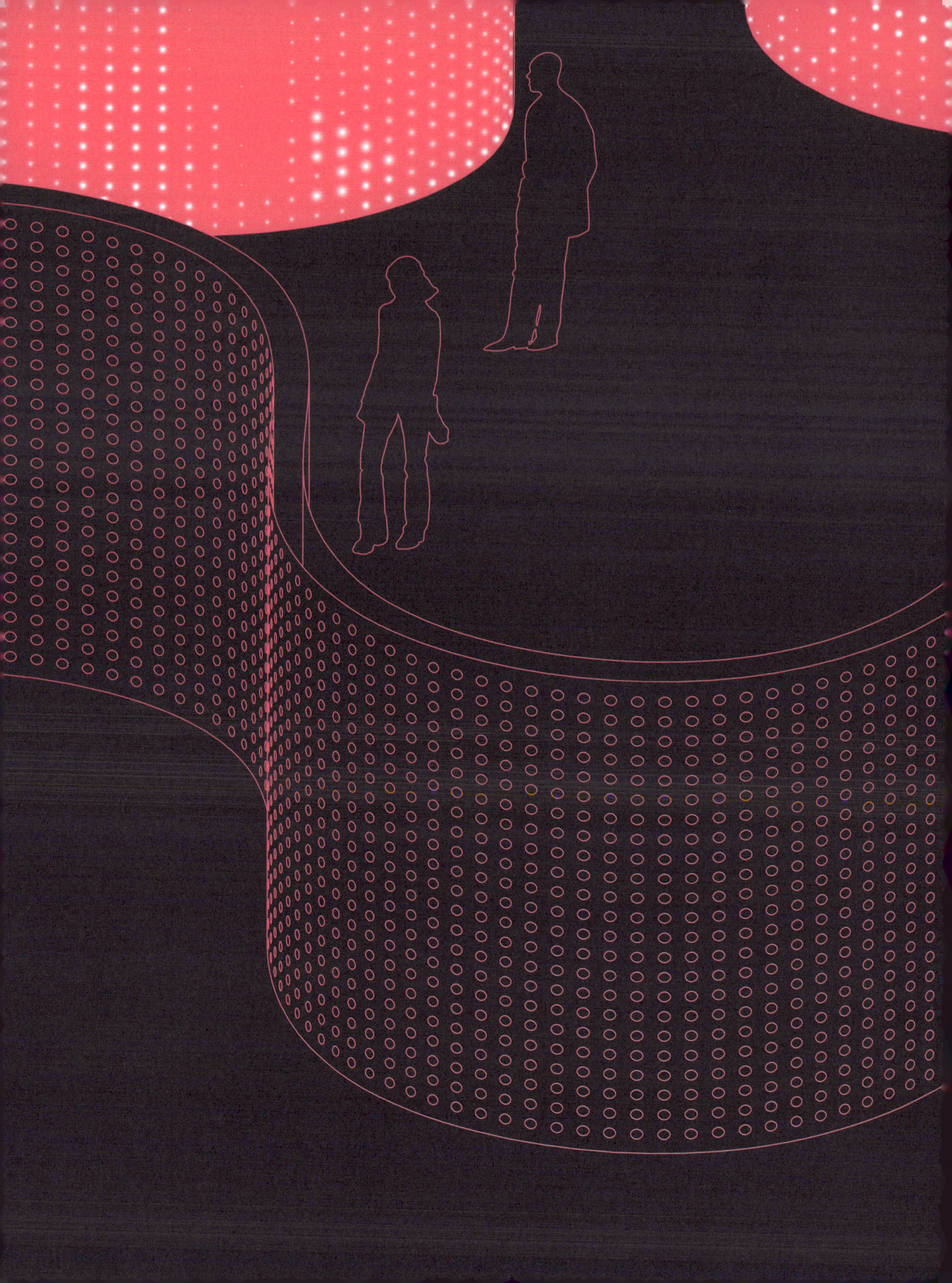

06
co2
07
10
11

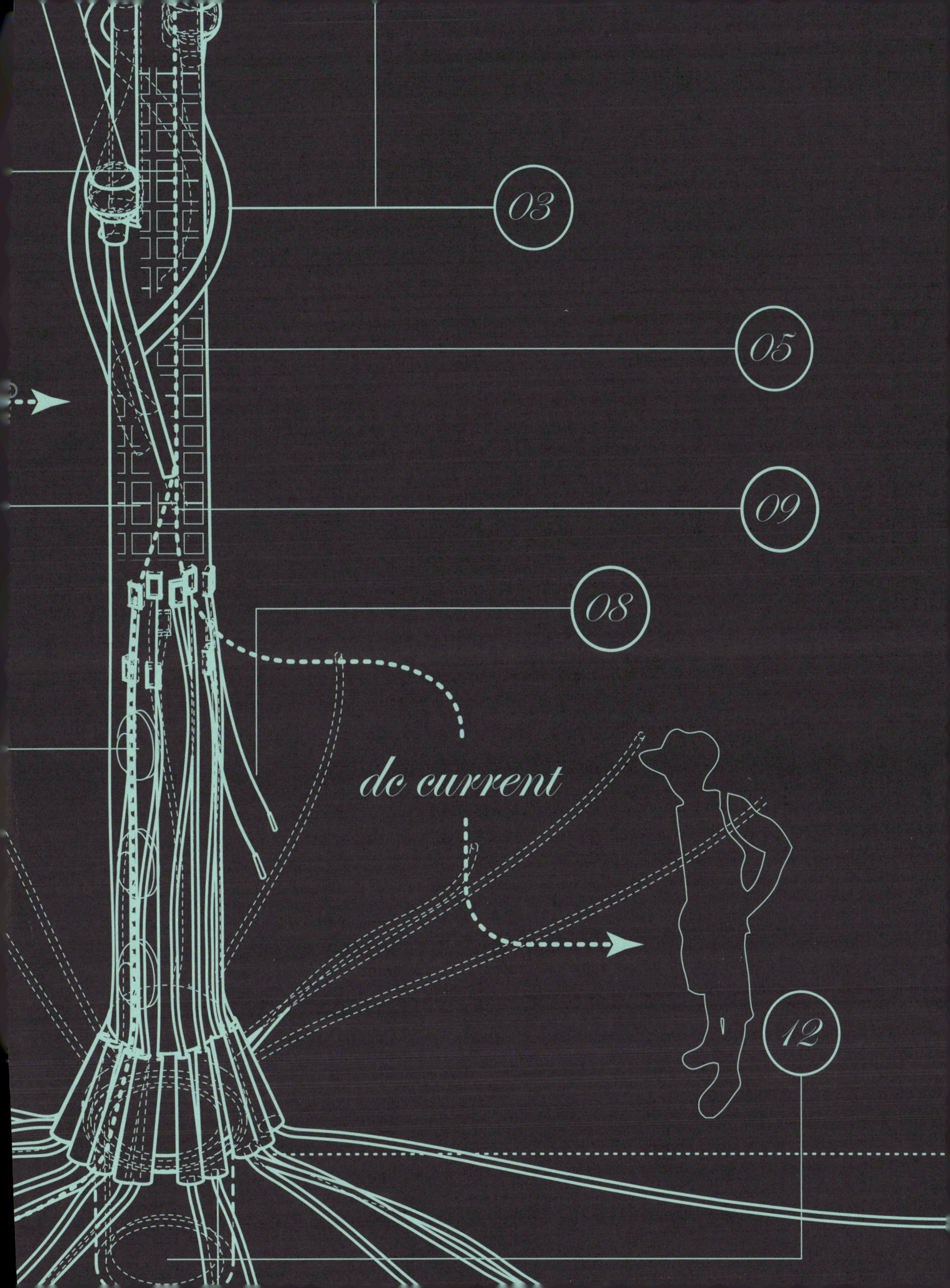
03
05
09
08
dc current
12

Book Credits
Authors: Carol Moukheiber, Rodolphe el-Khoury, Christos Marcopoulos
Managing editor: Oscar Riera Ojeda
Creative Direction by Leo Malinow
Graphic Design by Alejandra Román
Copy Editing by Kit Maude
Cover & Flaps image by Justin Cheng

With contributions by Drew Adams, Yuliya Bentcheva, Cleo Buster, Justin Cheng, Maya Desai, James Dixon, Holly Jordan, Khoury Levit Fong, Samuel Ko, David Long, Melissa Lui, Ricky Luk , Mani Mani, Fadi Massoud, Valentina Mele, Taraneh Meshkani, Azadeh Mohammadi, Nashid Nabian, Maggie Nelson, Otto Ng, Ada Juwah, Shadi Ramos, Julianna Sassaman, Mae Shaban, Don Shillingburg, Rick Sole, Lailee Soleimani, Studio NMinusOne, Elaine Tong, Jimmy Tran, Arthur Tseng, Nadya Volicer, and Ali Yarbakhti

—

The projects featured in this collection are authored, co-authored, or guided by Rodolphe el-Khoury, Christos Marcopoulos and Carol Moukheiber in the context of the following offices and institutions: Khoury Levit Fong, MIT, Studio NMinusOne, University of Hong Kong, University of Toronto.

—

The sampled projects represent the nature, scope and ambition of research at RAD. The lab that was set up and launched by the authors in 2011 at the University of Toronto to explore the future of embedded technology in buildings, cities and landscapes.

ISBN 978-988-16194-6-4
Printed in China

Oscar Riera Ojeda Publishers Limited
Unit 1A 20/F.,
Far East Consortium Bldg.
121 Des Voeux Rd Central, Hong Kong.
T: +852-5311-1625

Editorial Offices | USA
143 South 2ND Street, Suite 208
Philadelphia PA 19106-3073

Production Offices | China
Park A, Fang Da Cheng, Longjing Belhuan Avenue,
Nanshan District, Shenzhen 518055
T:+86-182-0098-3774

www.oscarrieraojeda.com | oscar@oscarrieraojeda.com